Mind Hacking

How To Rewire Your Brain To Stop Overthinking, Create Better Habits And Realize Your Life Goals

Jennifer Ferguson

Copyright 2019 © Jennifer Ferguson

All rights reserved.

No part of this guide may be reproduced in any form without permission in writing from the publisher except in the case of review.

Legal & Disclaimer

The following document is reproduced below with the goal of providing information that is as accurate and reliable as possible.

This declaration is deemed fair and valid by both the American Bar Association and the Committee of Publishers Association and is legally binding throughout the United States.

Furthermore, the transmission, duplication or reproduction of any of the following work including specific information will be considered an illegal act irrespective of if it is done electronically or in print. This extends to creating a secondary or tertiary copy of the work or a recorded copy and is only allowed with an express written consent from the Publisher. All additional right reserved.

The information in the following pages is broadly considered to be a truthful and accurate account of facts, and as such any inattention, use or misuse of the information in question by the reader will render any resulting actions solely under their purview. There are no scenarios in which the publisher or the original author of this work can be in any fashion deemed liable for any hardship or damages that may befall them after undertaking information described herein.

Additionally, the information in the following pages is intended only for informational purposes and should thus be thought of as universal. As befitting its nature, it is presented without assurance regarding its prolonged validity or interim quality. Trademarks that are mentioned are done without written consent and can in no way be considered an endorsement from the trademark holder.

Table of Contents

Introduction 6
Chapter 1: What Is Mind Hacking? 15
Chapter 2: Emotion and Your Brain................30
Chapter 3: How Emotion Translates to Behavior: The Good, the Bad and the Ugly 49
Chapter 4: Neuroplasticity and the Science Behind Forming Habits 72
Chapter 5: Letting Go of Worry, Overthinking, and Anxiety...............................91
Chapter 6: Mind Hacking Step 1: Identify Negative Influences and Habits...................... 111
Chapter 7: Mind Hacking Step 2: Set Your Goals and Learn to Visualize 130
Chapter 8: Mind Hacking Step 3: One Step at a Time.. 150
Chapter 9: Meditation Techniques 170
Chapter 10: Moving Forward: 10 Daily Practices to Strengthen Self-Confidence 190
Conclusion... 212

Introduction

A few years ago, studies were released with an alarming set of data regarding human beings and technology and how that relationship is responsible for decreasing the average person's attention span to around 8 seconds. We've all heard similar statements in the media or in news articles about how technology is changing our brains, how we've become addicted to our phones, and how this affects one-on-one personal interaction.

This is just one of the many reasons why it has become necessary to essentially "reboot" our brains in order to clear away all that excess information with which we are inundated each and every day. There is actually a phrase associated with this—information overload. Information overload is a product of the fast-paced nonstop society we live in where every electronic device, billboard and building is covered in advertisements, questionable news stories, celebrity gossip, and an endless array of entertaining videos, quizzes, pictures, gifs, etc. It is impossible, unless one completely unplugs, to avoid this onslaught. Unfortunately, many of us are tied to our technology for various reasons, and setting aside our devices for more than several hours is something many can't handle without worrying about family, feeling anxiety about missing texts, or missing important communications from our bosses or clients.

As a result, many people are drug into the daily cyclone of information overload which leads to added stress, anxiety and lethargy at the end of the day. Though stress can come from a variety of other factors in our lives, technology and information overload is usually near the top of the list, whether we realize it or not.

Add to this pressure at work, family or relationship issues, fitting in time for friends and social obligations, and caring for children, and you begin to cultivate a perfect environment for too much stress that often leads to anxiety and even depression. Stress is also a contributing factor in many chronic diseases, including those which affect cardiovascular health. The pressure to keep up and be successful is compounded in our society by the social media we feed our brains every day which show people at their best, happiest and most successful on a daily basis. It is natural to look at people like this and feel that we are in some way not doing enough, not making enough money, not working out enough, not enough of *anything*. This is where the low self-esteem, anxiety and depression can sneak in. So, if the stress isn't getting to you, the constant barrage of media telling you to do better in nearly every aspect of your life surely will.

What are we supposed to do? Well, the response to the kind of society we live in today is reflected in the myriad strategies that have exploded onto the market place. Everything from yoga, dieting and exercise strategies to energy drinks, vitamins and motivational conferences have been marketed as a potential solution to everyone's problems. This product reduces stress, that product makes you feel good about yourself, this dress will make you feel more confident...the list is unending as marketers continue to find ways to convince you that if you just buy their products, all of your problems will go away. So why isn't everyone perfect, beautiful, happy and healthy?

The problem here is that people keep trying to improve on their perceived imperfections by buying material things meant to solve their problems. There is a reason the adage "you can't buy happiness" has been around for so long.

When people throw money at their problems, the problems don't actually go anywhere, they're just covered up. This is not solving, it's camouflaging. As soon as that product has run its course, the problem will remain right where it was before. I'm not saying taking your family out to the movies to decompress from a stressful week is incorrect. What I'm saying is that using this or other temporary fixes to run away from constant stress that is taking a toll on your body is not the best way to address it in a meaningful and long-lasting way.

All of these quick fixes are external. Marketers and advertisements are creating a desire within you for their products and services as an external source of peace, calm, and happiness. But therein lies the crux of the problem—these things can only come in true form from *within* ourselves, not from the outside.

Everyone has goals for their lives. That goal may simply be to figure out what makes you happy, what fulfills you, or what fills you with purpose. Nobody other than you can determine these personal goals, and nobody outside of yourself can make you realize them. Sure, people can give you advice and show you the way, but it will be up to you to get there and reach those goals for yourself. You can hire all the personal trainers in the world, but unless you do the work yourself, you won't be seeing those physical improvements you've always wanted.

Similarly, the strategies and tips offered in *Mind Hacking* will be up to you to implement and incorporate into your life. Happiness and joy are personal, meaningful experiences that can only come from within yourself. It is what you do and how you think that ultimately determines your long-term mindset for your life—no product can just hand it to you. This book will challenge you to look deep inside yourself to address the thought patterns that are holding you back.

In Chapter 1, we will look a little closer at mind hacking and what we actually mean when we use the term. How do you hack your own mind? Can you actually change your brain?

Next, in Chapter 2, we will dig in a little deeper into the science behind emotion and your brain, how the brain connects emotion with experience, and how this can dictate how you respond to similar experiences in the future. We will learn how your emotions translate to behavior in Chapter 3.

Chapter 4 will explain the scientifically proven phenomenon that is neuroplasticity. This concept is why it is absolutely possible for you to, literally, change your brain. We will discuss this phenomenon using specific examples, and you will learn how this can be applied in your own life.

In Chapter 5, we will address some of the most common afflictions in our modern-day society, including excessive worrying, overthinking and anxiety.

In Chapter 6, 7 and 8, you will learn step-by-step how to begin changing your brain through consistent, daily practice. You will first learn to identify your personal goals, remove the negative influence and clutter from your path, then take your first steps forward toward developing new habits and realizing your personal goals, whatever they may be. It is important to note that this step-by-step guide can be applied to pretty much any goal you have in life.

In Chapter 9, we will introduce some meditation techniques that have been proven to sharpen the mind and help focus you in on your goals. There is nothing better than a daily meditation to help the brain clear away extra information, repair and refocus.

Finally, in Chapter 10, we will go over 10 daily practices designed to further boost your chance at success through strengthening self-confidence. You must believe in yourself and your own capability in order to keep moving forward. We know you can do it, and we are excited to share this journey and the secrets to mind hacking with you. So, let's get started!

Chapter 1: What Is Mind Hacking?

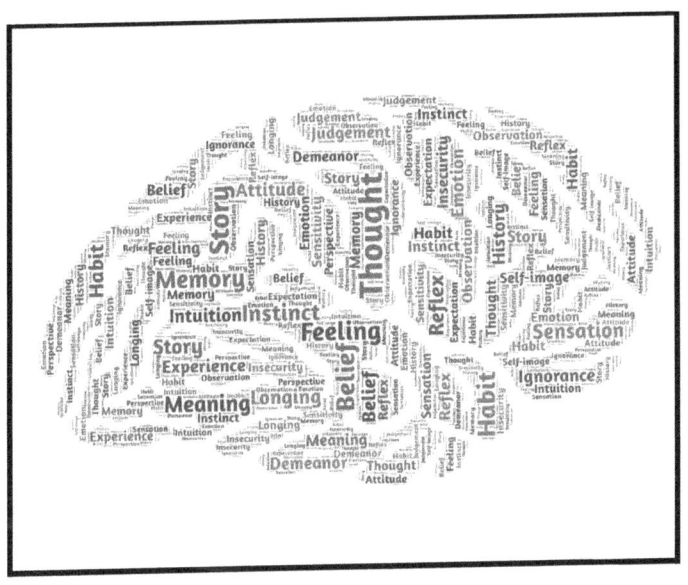

You've heard of hacking into a computer or hacking into a security system. Usually it's a smart protagonist or a criminal in a TV show whose fingers effortlessly move over the keys with a complex series of numbers on a computer screen before she hears a click or another sound to signify that she's been successful. So, what can we possibly mean when we talk about mind hacking? Is there a similar security system in place for which there is a magic code to get in? The analogy is an interesting starting point and has become a popular catch phrase in the media, but I'm going to explain to you what exactly I mean when I write mind hacking.

To do this, I'd like to illustrate with some examples of how the brain functions when it is allowed to simply "go with the flow" in our modern-day society.

Groupthink

A young man named Simon attends a staff meeting at work. He is part of a group of 7 people working together on a new software program for their tech company. The personalities of the people in the group vary. There are a few shyer individuals who tend to let others do the talking and prefer to work alone. There are a few individuals who actively participate in meetings and like group work. Then there is one man, Jim, who tends to take the lead in conversations. It seems to happen naturally, and he's not trying to take control from others. He is simply the most outspoken person on the team and very invested in his work. As a result, he usually has strong opinions about how to move forward on a project.

When he speaks, he uses a strong, loud voice that everyone can hear, and it sounds as though he really knows what he is talking about. When others in the group speak, they may feel just as strongly about their own opinions, but when it comes time to vote on a decision for how to move forward, Jim's plan usually wins. He's appealed to the basic tendencies of people to move toward those who display strength and leadership, even independently from what they think of this dominant person's words. Without thinking too deeply about Jim's words, the group is already immediately swayed toward him because of the conviction in his voice and the authority of his tone. Perhaps Simon introduced an opposing thought, but he spoke more quietly and less authoritatively. His idea may well have been something to consider for the project, but because he came off as less confident, the group gravitated toward Jim's ideas instead.

This is one example of a phenomenon called "groupthink," though groupthink encompasses a much wider arena of human behavior. Individuals moving together toward a singular, authoritative figure for the sake of harmony and group solidarity is a human tendency that has been around for thousands of years. A more familiar phrase you've probably heard many times is "peer pressure," which refers to the direct pressure one receives to behave the same way as the rest of a group. Peer pressure is very strong in the years of a person's adolescence because this is the time when he is trying to figure himself out and striving to find others with whom he can relate and experience this struggle. People will do a lot of surprising things for the sake of conformity and retention of status. It is hardwired into our brains to cultivate some kind of social status in order to be protected by the group. When you go against the grain and move in your own circle as opposed to others', you set yourself apart and become unimportant to those who choose to conform with their own. This was dangerous in the days when human survival depended on the strength of one's community as a unit for hunting, gathering and surviving.

The pressure to conform starts very early in life and continues throughout one's lifetime. A person will experience pressure from peers, family and authority figures insisting on exerting their influence on his/her behavior, thoughts, and belief systems. In addition, the person further experiences pressure to conform culturally through media and marketing efforts, as we've already discussed.

These influences have a much higher degree of success in winning over people who are not very self-aware. Some people simply do what they are told throughout their lifetimes. They may not spend very much time on who they really are, what is meaningful to them, or what their true belief systems are. They seem content to glide through life at someone else's direction, and these are the people who often make easy targets for those who wish to take advantage of their impressionable characteristics.

Cult mentality is also a good example of the phenomenon of groupthink at work. And it would be a mistake to think that only "weak-minded" people are susceptible to buying in to a cult or a cult leader's ideas. Victims who have spoken about their experiences talk of members who espouse a clear methodology when it comes to recruiting new followers, and their targets are weaknesses which are present in any human being who has experienced pain and adversity in some form. To demonstrate, think for a few seconds about a painful emotion that is connected to a memory. Something that has haunted you for years and which you've tried many times to run away or hide from. Now imagine someone coming up to you and sharing with you a similar experience she has had. Then, she tells you that she's found a way to permanently heal that wound and fully recover from that emotion. Would you keep listening?

Few of us wouldn't. Scam artists in many forms target and profit from their claims of being able to heal painful wounds. And this is just one reason why it is necessary to strengthen your mind, become self-aware and cultivate your inner confidence and conviction that you can change and mold your life according to your own personal goals; not anyone else's.

Do not adopt others' goals and ambitions for you if they are not what you believe would make you happy or fulfilled. It's hard to combat this influence if you haven't yet sat down and really thought about what these goals for your life actually are. Many figure it out too late, as they travel down a long road only to discover that they don't understand why they've come that way.

Mind Hacking Exercise #1

Mind hacking is about breaking free from the zombie-mode way of life. It is about stopping in your tracks, taking a look at yourself and your surroundings, and manually changing the trajectory of your life. A good starting point for you might be to think about and consider why you picked up this book. Why are you still reading? Has something you've read struck a chord within you? Does the scenario with Jim and Simon sound familiar to you? Do you also feel like you are following someone else's plan for your life?

Perhaps you've already discovered that there are things and thought patterns holding you back from progressing in your life in the way that you want. You are ready to proactively make a change and get rid of those habits and negative influences that keep you waking up stuck in the same spot every single day. Whatever your situation, I promise that by the end of this book, you will feel empowered and prepared to take hold of your own life and begin steering yourself toward the future you strive for.

So, let's find a good starting point for you before diving into the science behind emotions, behavior and neuroplasticity. It is important to feel oriented at the beginning of your journey before we move forward so that you don't feel lost or confused later. Set aside some time today to think about yourself and your life. I would strongly encourage writing down your thoughts in a journal and that you continue to record your thoughts and progress in this journal as you incorporate the strategies and tips in this book into your life.

First, think about or write down what it is in your life that makes you most happy. What makes you sad? Do you feel you spend a lot of time unhappy? How about anxious, depressed, or stressed out? Do you feel trapped, as if you have no choice in whether or not you feel these emotions? Before trying to move forward or adopt strategies, you have to understand where you are in life and why you are unhappy. Is there something missing from your life? Remember, happiness doesn't come from things, and it doesn't even necessarily come from other people. Happiness must be cultivated from within yourself. But perhaps there is something that you used to do as a child that made you happy that you no longer have time to do. Maybe you've really wanted to get involved with volunteering or a group that does weekly community service work as a way to give back. Perhaps you really just want to meet like-minded people who are in to the same hobbies and interests as you. Whatever it may be, now is the time to home in on the specifics and write them down.

Perhaps your happiness stems from the idea of removing some kind of negative influence or habit from your life. Perhaps you've tried for years to quit smoking and have been unsuccessful. Or maybe you have a friend who keeps encouraging you to be irresponsible and you think he/she might be a negative influence holding you back from achieving your goals. Write these down as well. In order to fully orient yourself, you must have the clearest picture possible of where you see yourself when you've achieved your life goals. You must also have a clear picture of the things which it might be time to let go of.

Don't worry about nailing all this down immediately. It will usually take a little time and reflection to come to terms with what needs to change in your life. As you discover new insight or come to decisions about where you want your life to go, write them down so that you can remember them and come back to them regularly.

Once you've gotten to this point, I'd like to introduce your first mind hacking exercise. Are you ready?

Find a quiet place to yourself where you are pretty sure you won't be interrupted. We're going to practice a simple form of mindfulness meditation to get you focusing in on where your thoughts tend to wander. All I want you to do first is observe your thoughts. Where does your mind automatically wander? Do you feel silly for trying the exercise? Maybe you keep thinking about something awkward that happened between you and a coworker almost a week ago now.

Now observe what kind of emotions you are feeling. Are they associated with some of these thoughts? Do you feel a wave of embarrassment when you think about something that happened at work recently? Do you feel residual anger from the traffic jam you had to suffer through on the way home from work? Don't push yourself to think of every single detail or to pinpoint your thoughts exactly and define them with emotions. Perhaps you are relatively calm and feel at peace just sitting and paying attention to your thought patterns. This is the idea.

Now, pay attention to your breathing. Breathe in for 5 seconds, hold for 5 seconds, then release the breath for 5 seconds. Don't try to push away your thoughts or think of "nothing." The idea is to simply take note of your breath and try to observe your breath for several seconds as you breathe in and out.

Congratulations! You've completed your first step toward hacking your mind. As we continue, you'll understand more about the significance of becoming self-aware as you create a new design for your life.

Chapter 2: Emotion and Your Brain

When we talk about emotion, most of us feel like we have a pretty good understanding of the concept. Emotion is what we feel when we are in love, and when we feel sad, angry, or afraid. These are kind of the big four categories of emotion that most of us think of, and most other nuances of emotion fall under one of these umbrellas. What you might not be familiar with is how exactly these emotions manifest in the body.

On the simplest level, our brains are constantly looking for two main things—danger and reward. When the brain detects danger, it sends out chemicals that make us respond. This is what's referred to as the "fight or flight" response. For example, if a person is out in the wild and they hear a sound that they've heard before in connection with a dangerous predatory animal, then he will become frightened. He will then choose between fighting and running away based on his knowledge and past experiences.

Most of us don't experience the exact same threat in modern-day life, but our brains still develop with this hardwiring that prompts us to respond in this way when we perceive danger. The brain releases what are called stress hormones—adrenalin and cortisol. These are responsible for the surge of energy we get when we are suddenly frightened. You may have heard stories about humans adopting a kind of super strength in times of acute stress. This is the adrenalin pumping through their bodies making them capable of dealing with the situation.

Similarly, when the brain detects a reward, it releases chemicals that make us feel good, including dopamine, oxytocin, and serotonin. These signal to the body to continue with whatever activity is activating the pleasure receptors. When we exercise, we get a rush of good feelings that encourages us to continue the activity regularly. Eating has a similar effect.

Modern Emotional Addiction

The trouble is that in today's world, we've come a long way from those times when the chemicals in our bodies signaled danger and pleasure to increase our chance for survival. Nowadays, most people on Earth don't have concerns in the same form and it is easy to abuse what makes us feel good. For example, eating is necessary for life, of course, but the brain can become addicted to the pleasure of eating food. Especially if we are eating unhealthy food, this addiction will start to affect other systems of the body and even deregulate the signal in our brains that signals fullness. When this happens, a person might overeat easily and not be able to tell when they are full, but the immediate pleasure remains the same. The response is deeply hardwired into our brains. Pizza and cookies will continue to initiate a pleasure response, even if we take them in at an unhealthy rate.

Illegal drugs form addictions in their users because they go right to these pleasure receptors and flood the body with pleasure. If a drug is injected, the effect is almost immediate. The person's need for the drug then escalates because he/she begins to need more and more of the drug to reach the same high as when they first started using. It is a vicious cycle and one that is very difficult to escape.

Did you know that we can also form addictions to our emotional responses? Let's look at some examples.

Do you know anyone who you would characterize as an "angry" person? Perhaps this is because he/she always seems to be in a bad mood when you see him/her. Or maybe you know someone who just seems to constantly be the victim of bad luck and is constantly whining about his life and the things that happen to him. The fact is, people can also become addicted to cycles of emotion.

To simplify this notion, think of the brain as being split into the "feeling" brain and the "thinking" brain. The feeling brain is in charge of these emotion responses to nuances of fear and pleasure. The thinking brain is in charge of our actual thought processes and reasoning. The fact is, the feeling brain kicks in quicker than the thinking brain, so it is easy to have an emotional reaction to an event before we can even rationalize what is happening. Let's examine the above examples a little further.

There is a man in your office named Sam who just always seems angry. He gets upset at the littlest incidences and seems to carry this emotion around with him all day. The anger response is triggering because he's trained his mind that this is the appropriate response to events which inconvenience him in some way. Now, it may have started out small, like getting stuck in traffic and being frustrated about it. But the longer Sam responds to things like a spilled cup of coffee, someone coming in late to work, missing a show on TV he was really excited about, having an argument with his wife, etc., with irrational frustration before engaging his thinking brain, the more his brain adopts this as the normal and expected response mechanism.

Let's imagine another man in the office whose name is Tim. Tim tends to come in to the office with his head downturned. He just always looks defeated in some way and is always complaining about things that happen to him on a daily basis. He feels the world and the people in it are just unfair and terrible and that he will never be able to improve his life. How did he get this way?

Well, he's formed an automatic response system to things in his life which he feels he cannot control. The process begins at the very beginning of the day and continues throughout the rest of the day until he is ready to go to bed. This is an extreme example, but the gist of the situation is that he's also taught his brain to respond with sadness and defeatism whenever something happens that sets him back in some way. These events could be as simple as getting a bit of his lunch on his shirt during his lunch break. His brain has been trained to respond with an overabundance of negative feelings about this, and thus Tim responds to the mishap as if someone had just run over his dog. He, too, has developed this emotional addiction over time as he continued to cultivate a general feeling of defeatism in response to the world. It has likely affected many other aspects of his character, like self-esteem and confidence.

Emotion can be an incredibly destructive thing if we let it form addictions in connection to our brain's response to life events because our brain also influences how we see others and their emotions. It's a strange phenomenon, but very true.

Let's imagine a woman named Valerie who experienced a traumatic breakup after her husband cheated on her in her early 20s. The experience prompted a serious depression and thereafter she's harbored an intense feeling of suspicion and dislike around men in general.

In this case, Valerie has connected her emotional experience with one man to all men she meets afterward, and this emotional response may continue for years and even for the rest of her life if it is not addressed. Her brain has made actual connections over time that link men and feelings of fear, anxiety and sadness, limiting the possibility that she will find a satisfying relationship with another man.

The influence of this emotional experience may go even further. To illustrate, let's imagine Valerie does finally meet someone with whom she's willing to try a new relationship. When Valerie is out in public with her new boyfriend, let's call him Zack, she reads things into his face that may not be accurate based on her past experience and assumption that all men are looking to cheat. If they pass an attractive woman on the street, Valerie may look over at Zack and believe she's seeing a look of desire on his face. When she confronts her new boyfriend about it, he denies that he was checking out this woman, much less desiring her right in front of Valerie. This is going to be very difficult for Valerie to accept because in her mind, she saw *with her own eyes*, the emotion of lust on his face.

Many people hold the assumption that emotion itself is connected in some way to a universal system of facial gestures and expressions, and this just isn't true. We may be able to make informed guesses about someone's emotions based on their facial expressions, but the truth is, the emotions we read in other people and on their faces largely come from inside our own brains, just like every other representation of our world. We all see this world differently.

Research has been done to demonstrate this phenomenon. People are given a picture of a man with what would generally be considered a completely flat affect and facial position. Volunteers are asked to fill out a survey about their own emotional states, past trauma, how they are feeling, etc., then they are asked to read the emotion on this person's face. What did they find? People's reaction and evaluation of the photo reflected the emotional states they'd reported on their self-evaluation forms. Those who had experienced recent trauma or sad events reported reading sadness in the face of the man in the photo. Those who reported a generally good attitude with no recent misfortune said they could see a hint of a grin on the man's face. Others reported anger or fear. The results are startling and tell us a great deal about our human tendency to "read in" to our daily lives and the people involved in them.

So, what does this mean for you? Are you trapped forever if you've found you have a vicious emotional addiction that pops up every time you hit a red traffic light? Well, I have good news.

Once we understand that our emotions tend to hijack our brains before our thinking minds can rationally respond, we can create a plan of action to slowly begin to adjust these emotional reactions. Research shows that we can absolutely influence our emotional responses through changing our thought processes, and this is part of what mind hacking is all about.

Interruption of Thought Patterns

The first step to address an adverse emotional addiction is understanding the emotional response. Where do you believe the emotion is rooted? Were you influenced by someone else in your life who tends to react a certain way? Perhaps you've experienced something traumatic that has influenced your outlook on life in general. Whatever the root cause, it is important to really think about why you react a certain way when it is something you very much want to change. It is the same principle as the prior exercises we did for discovering where your mind tends to focus throughout the day and writing down life goals. You can't start moving toward a goal without first having a clear idea of what it is, but also of what's blocking your progress.

I suggest you get out that journal again for this step in the process, as writing is usually a good way to really flesh out the detail of emotional responses. Write down your problem emotion and then write as much as you can about why it is you think you tend to respond that way, as well as why you think it is irrational. To be clear, we're not trying to eradicate sadness or fear or some other painful emotion from our lives. These emotions are relevant and incredibly important for a healthy human being. What we want to address are those cycles of emotions that do more harm than good and which we recognize as out of proportion with the stimuli. In other words, getting upset after stubbing your toe is reasonable; throwing your coffee thermos across the room every time someone does anything remotely annoying is probably not so reasonable!

I want you to look at your thoughts after you've finished outlining one adverse emotional cycle.

Perhaps you have more than one, but don't overwhelm yourself. Focus in on one at a time.

Now, I want you to conjure an imaginary situation in your mind in which you are experiencing the opposite, positive emotion in opposition to this negative one. If you struggle with fear, imagine yourself in a situation where you are exuding bravery. If you struggle with anger, imagine yourself in a happy, stress-free situation, etc. Essentially, what we are doing here is finding a thought pattern to counteract the ones which follow a negative emotion. So, if you feel a blow to your self-esteem every time you see that one woman at work who seems to have the perfect life, imagine yourself walking in to work confident and with a smile on your face. You've achieved certain goals in your life, and you are feeling good about it. This is your weapon, and you will need to take some time to visualize and feel the positive emotion that comes with this imaginary scene.

Now, the next time you feel that negative emotion being triggered, focus your mind on that positive, counteracting thought and see what happens. It may take some practice, but it is important to be patient with yourself. A habit is never easy to break, and it will take consistency and a dedication to the change. Over time, though, if you can implement this weapon in the majority of cases when your negative response usually takes over, you will see a change in your behavior and overall mood. This is because you are actually rewiring your brain to work differently in those situations. Don't take my word for it. Try for yourself. Keep a record of your progress in your journal so that you can look back and see how far you've come. This will serve as a strong source of confidence and motivation as you move forward.

I'd like you to try one more simple exercise that you can do anywhere you feel comfortable. This is an exercise I've seen many motivational speakers across the world use in their workshops to demonstrate the power of the mind over emotion. Find a private spot to conduct your experiment, or if you are feeling confident, try the exercise out at the park. It might work even better!

There is an additional element that works hand-in-hand with thought exercises when it comes to altering emotion and self-confidence, and that is physical movement and posture. Did you know that simply forcing yourself to smile even when you don't feel like it will raise your mood? Try it!

Sit down comfortably or stand up straight and look straight ahead. Now, without conjuring any specific thought pattern, simply move your mouth to form a big smile. Hold it there for a few seconds. How do you feel? Do you feel a change? Did you feel silly and start to laugh? Laughter is a very positive emotion, you know.

Now, let's conduct a posture experiment. Stand up straight wherever you are with your shoulders back. Think of a time in your life when you achieved a goal or accomplished something that was a really big deal to you. Something you consider one of the greatest achievements of your life so far. Now, raise your arms up in a victory pose with your arms making a "V" above your head and your hands forming fists. Hold this pose for a few seconds. How do you feel? Did your mood change at all? Do you feel more confident? Most people respond with a surprised look on their faces as they say, "yes!"

Now, see if you can hold this posture and alter your mood. Try your hardest to start feeling sad or to lower your confidence level. Hard to do, right? Our body posture has an influence over our emotions, just like active thinking does.

Assume the opposite position with your head downturned and your shoulders stooped. Think of a time when you failed miserably at something. This should bring down your mood slightly. When you try to change your confidence level in this pose, it is also very hard to do. Your posture is so suggestive over your brain that even when you try hard, it is difficult to feel confident and happy while holding this abysmal posture.

We've looked at all of these examples in order to illustrate the malleability of emotional response. You do have more control than you think, and you can change your thought patterns for the better. In the next chapter, we will look at how emotion translates into human behavior.

Chapter 3: How Emotion Translates to Behavior: The Good, the Bad and the Ugly

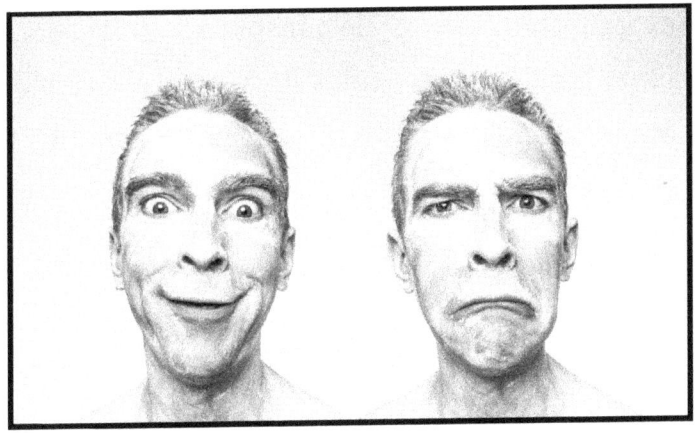

In this chapter, we are going to look at a few of the most famous and controversial social experiments ever conducted in order to gain a better understanding of how emotions translate into behavior. The subheading of this chapter is, of course, taken from the classic Clint Eastwood film, and it will seem like we are mostly focusing on the bad and the ugly parts of human psychology, but we will end the chapter on a positive note. Just like we discussed in the last chapter, it is possible to rewire your brain to get off autopilot. These experiments have demonstrated just how important it is to cultivate a sense of self-awareness and personal principle, even if it means going against the grain of how others behave.

The Marshmallow Experiment

The marshmallow experiment was first conducted by experimenters at Stanford University in the early 70s and involved a group of young children. The children were brought in and asked to sit at a table with a big delicious-looking marshmallow sitting in front of them on a plate. The experimenter explained to the children that he was going to leave the room, and if the children could wait patiently a full 15 minutes without eating the marshmallow, they would get 2 marshmallows as a reward when he came back.

When the experimenter left the room, some of the children immediately picked up the marshmallow in front of them and ate. Others squirmed around and tried to wait but ended up eating the marshmallow just a few minutes later. Finally, a few of the children did manage to wait the full 15 minutes and receive the reward when the experimenter came back in the room.

This may not seem extraordinary by itself, but the most interesting discoveries were made long after the initial experiment. The children were observed from afar as they grew up and into their adulthoods. There ended up being an incredible connection between whether or not the child could wait to eat the marshmallow and the success they experienced later in life. Those who were willing to wait experienced more success in their lives than those who were not patient enough to wait before eating the marshmallows. The experiment offered a great deal of insight connecting patience and success over time.

The difference in these children was the natural propensity and capacity to understand effort in exchange for reward as well as patience. Being able to wait patiently allows one time to think through situations before acting too fast, and when this pattern of behavior is reinforced over time in the brain, it becomes natural and habitual to respond to situations by first pausing to think things through. Those who demonstrated a lack of patience and self-control were more susceptible to cultivating a pattern of behavior that allowed for acting first and thinking later. Without conscious effort to practice patience in life, a person is more likely to make rash decisions without thinking them through first. These individuals may also act on their emotions much more quickly and intensely, leading to unpleasant patterns of emotional triggers throughout one's lifetime.

Milgram Experiment

Let's move on to a much more disturbing and controversial social experiment conducted by a Yale professor in the early 60s. It is said that the experiment was born out of a curiosity that was prompted by a recent event in which a former Nazi defended himself and his actions in accordance with the Nazi party by saying that he was simply "following orders." This got Professor Milgram thinking. What actually happens when a person is put in a situation in which there is an authority figure applying pressure to follow orders in accordance with an established system or in harmony with others in a group of followers already following the orders? Out of this curiosity, Milgram created the now notorious Milgram experiment.

Subjects were brought into a room where a device was set up that was designed, they were told, to administer electric shocks at the flip of a switch. Another participant, whom the subject believed was also part of the experiment, was strapped in to a chair and connected to the device so that when someone flipped the switches on the other side of the wall, he would feel electric shocks equal to the number of volts above each switch...or so the subject believed.

What the subjects in the experiment did not realize was that the device designed to administer electric shocks was completely fake, as was the concept that flipping these switches would hurt the man strapped into the chair in the adjoining room. This man was an actor hired for the purposes of the experiment.

Subjects listened to their directions intently, though not without a bit of surprise and hesitation. The experimenter explained that the subject would sit in front of the electric shock device as the subject communicated with the other subject strapped in and connected to the device in the opposite room. He was to read off a problem for the man to solve. If the man answered the question correctly, nothing would happen. But if the man answered incorrectly, the volunteer was to administer an electric shock as punishment. Volunteers were told that the experiment was all about the utility of punishment in the process of learning and were unaware that they themselves were actually the real subjects of the study.

The study got interesting as the experimenter explained to the subjects that each time the man in the other room got an answer wrong, he was to up the voltage on the next electric shock administered. The volunteers were to go through a large list of questions and were asked to continue upping the voltage on the electric shock punishments. They were not given a limit or a point at which they should stop. At an interval of every few levels of voltage, there was a clear label identifying the voltage intensity, such as "minor shock" all the way to "danger severe shock," etc. The range of voltage went from 15 to 450 volts.

The questions Milgram wanted answered were: Just how far would people go? and why? The results were quite astounding.

As the volunteer sat down to begin, he would read off his questions and administer the shocks as prescribed when the "learner" got an answer wrong. This continued on up the scale of voltage until at a certain point, the learner began calling out in distress and pleading for the experiment to stop. He would yell things like "Let me out! Let me out! You can't hold me here!" Oftentimes, the "teacher" and subject of study would turn around to look at the experiment leader as if to ask, "Are you sure we need to continue?" Sometimes the subjects would verbally object to moving forward, but the experimenter would then apply a bit of pressure. He would say things like, "It is absolutely essential that we continue." After the statement, the experimenter would simply return to what he was doing with an absolute assumption that the experiment would continue. Faced with this pressure from a person they perceived as an authority figure and leader, a startling number of test subjects *actually continued* to lethal levels of shock. A whopping 65% of the participants, or two-thirds, actually went on until they'd administered the final shock of 450 volts. The *majority* of people who participated in this study were actually persuaded to kill another human being through what could be termed relatively minor authoritative pressure. What does this say about us as human beings?

We talked a bit about the ancient hardwiring in our brains in chapter 1 to maintain social status through adhering to social rules, often leading to the phenomenon of groupthink, when people gravitate toward the person exuding the highest authority. Similarly, the brain of the subject in this scenario is very quickly making decisions and forming associations which conform with what is comfortable and safe for them and their own status in the situation. The experimenter was seen as a strong authority figure with more knowledge and understanding of what was going on. In a position of doubt, the test subjects quickly identified themselves as the one with little understanding of the situation and deferred to the one in a position of power, knowledge and authority. Only in instances of extreme ethical opposition to the pain being inflicted on the learner were the test subjects able to stand up and withdraw from the experiment, their conscience outweighing their sense of self-preservation and appearances.

Out of these observations, Milgram and others who studied his work were able to analyze and postulate on how the emotional reactions in situations of intimidation, authority, peer pressure or all three at once influenced behavior. Human beings have an incredible capacity to rationalize and justify their behaviors out of fear or anxiety. Just as in the fight or flight response, the emotions triggered by the sense of impending danger essentially hijack the brain and prompt behaviors in advance of the thinking brain's assessment of the situation. Likewise, the threat of embarrassment or causing disappointment in the presence of a respected figure, and in conjunction with a very respected psychology department at a respected university, no less, was just too intimidating for participants to protest.

Perhaps this makes sense on this small-scale experiment level, but what about when we think about the atrocities committed by the Nazi party during World War II? Surely, no amount of peer pressure or authoritative influence is enough to justify the actions of those who participated in the Nazi atrocities, and that is absolutely correct. However, it would be a shame to ignore the scientifically relevant implications of this outrageous example of human behavior in the face of the Nazi party's influence. Much research and analysis points to a situation akin to the frog in a boiling pot of water analogy. If you've never heard of this analogy, I'm referring to the concept that you can actually set a frog to boil in a pot of water so gradually and so slowly that the frog will die without even realizing it was in danger. The frog adjusts its temperature according to the rising temperature of the water until just before boiling point, at which point it is unable to jump out because all its energy was given to adjusting to the water temperature.

Personal convictions, belief systems, and moral principles are all susceptible to outside influence and our emotional states. If we imagine a young man in the early stages of Nazi Germany, you have to understand the environment and atmosphere of the country and its attitude toward Adolf Hitler. Hitler portrayed himself as a very strong, confident leader, and his ideas about raising Germany up in status in the world was an alluring concept. We must see the gradualness of the change that would have taken place in those who became followers of the Nazi ideology as one idea is introduced at a time. After a while, a level of trust and confidence in one's leader is enough to feel an absolute sense of necessity to follow the leader's orders, even in situations where the individual may have previously questioned his actions.

A similar phenomenon happens within the cult mindset we discussed earlier in conjunction with groupthink. A conversion begins with a person latching on to one particular concept, then another idea that goes along with it, then another facet of the cult ideology. These ideas build upon one another, combined with an enthusiastic and charismatic leader, and soon you have a prior skeptic at home practicing cult rituals in the middle of nowhere as part of a notorious cult!

The Stanford Prison Experiment

Professor Philip Zimbardo's Stanford Prison Experiment from 1971 stands as perhaps one of the most outrageous and ethically questionable social experiments ever conducted. Nevertheless, we can't deny that the results of this prematurely halted experiment reflect a disturbing and fascinating part of human nature. The experiment truly altered the mental states and perspectives of the individuals involved, and to an alarming degree. Where there first had existed no great contrasts in general regarding socioeconomic status, disposition, or personal history, there soon appeared an alarming divide brought on by the simple introduction of power in a simulated scenario. One group of men were given the roles of the prisoners, and the others in the group were given the roles of prison guards. What followed over the next 6 days was truly extraordinary and was so notorious that a feature film was made depicting the incident in 2015.

A space inside the basement of a psychology department building was transformed into a small prison. There were bars on the cell windows and even a space set aside to serve as a place for solitary confinement, or "the hole." The guards were given "night sticks," sunglasses which covered the entirety of their eyes, and a uniform. The prisoners were brought in and were subject to strip searches before donning thin, flimsy prisoner garb for the duration of the experiment, which was to last two weeks.

On the first day of the experiment, nothing extraordinary happened. Everyone seemed to settle in to their roles with the knowledge that they were simply participating in an experiment.

The second day is when things started to get bumpy. The prisoners decided to rebel a little by barricading their doors with the beds they'd been assigned. The guards, in response, started to routinely punish the prisoners as a way to exert more authority over them. They'd been warned to do what they could to enforce order and maintain control, short of physical violence. Verbal abuse started to fly between prisoners and guards, and soon the divide between guard and prisoner was stark. Each of the individuals was starting to respond and take what was happening to them personally, and it seemed like they were gradually forgetting that they were participating in an experiment rather than a prison environment. One individual who was part of the prisoner group even decided that the experiment was too much for him, and he asked Zimbardo if he could leave. Zimbardo responded as a prison warden, explaining that he should consider staying and serving as an informant in exchange for better treatment. But the individual returned to the other prisoners with an understanding that he, in fact, was not allowed to leave, saying that no one else was going to be able to leave, either. Zimbardo maintains that he never told this individual he couldn't leave, but the prisoner had responded similarly to how a real prisoner would feel in this situation— powerless and at the mercy of the authority figures.

This perception fundamentally altered the experience for the prisoners, and relations between the two groups began to degrade further. Prisoners would be woken from their sleep to do disgusting chores, like cleaning toilets with their bare hands, and performing menial tasks and exercises. The process of degradation for the prisoners felt real, as did the feeling of power in the guards as they exercised their authority and control.

At one point, a prisoner who had disobeyed guards was punished by having him watch as his cellmates were punished for his behavior. They became angry with him, and the emotional stress had taken a toll to the extent that he asked to see the "warden" to ask to leave. Of course, Zimbardo told him he could leave, but down the hall he could hear his fellow cellmates chanting "Prisoner [prisoner number] did a bad thing!" over and over and over. In tears, this prisoner said that he couldn't leave and that he had to go back. He couldn't handle the other prisoners thinking he was a bad prisoner or a bad person. This really made Professor Zimbardo perk up because there had been this incredible transformation that had taken place within a short span of time. He had truly adopted this prisoner mentality, and what's more, he was feeling the pressure to conform within his own prisoner community. Zimbardo assured him, reminding him that it was only an experiment and that he wasn't a bad person or a bad prisoner. This helped clear his head, and the subject left the experiment.

The project escalated until a fellow psychiatry professor came to observe the experiment and ultimately convinced Zimbardo it was time to end it. Professor Zimbardo admits that he, too, fell into the mindset of playing the part of a prison warden and was caught up in the fascinating play of events happening before him.

The experiment was a game-changer, and subsequently there were guidelines and rules set in place to protect participants from experiencing such abuse and psychological pain as part of future experiments. But the harrowing results and implications remain and are still the subject of analysis and debate. The emotional feedback offered by power, in opposition to the emotional feedback in response to degradation, was powerful and long-lasting, according to participants who were interviewed after the fact. It speaks again to the susceptibility of everyday "normal" and ethical human beings to the influence of emotional response. Would the prisoners in the experiment have experienced such psychic pain had they continually reminded themselves that they were simply part of an experiment and that what was happening to them was not real?

Carlsberg Beer Experiment

I'd like to offer up one more example that may be a little more lighthearted to close this chapter on emotional influence on behavior.

As part of an ad campaign for Carlsberg beer, the company engineered a social experiment in which a theater for a movie playing one night would be completely filled with biker gang-looking guys, complete with leather jackets, gloves and beards...lots of beards. An unsuspecting couple would be led to the theater to watch the movie for which they'd just bought tickets, and they would enter a theater in which every seat was taken by these biker men except for two open seats right in the middle of the theater. Hidden cameras captured the reactions of these unsuspecting people. Some turned away as they associated these men with dangerous personalities and were uncomfortable. But those who simply walked up and took the two seats were rewarded with fanfare and free beer. (The looks on their faces are priceless, and you can view footage from the ad on YouTube.)

I want to take a moment to emphasize and augment the importance of the fact that, in the case of Milgram's study, there *were* participants whose consciences were able to help them rise above the influence of power and authority. It is not impossible for human beings to enforce their powers of logic and reasoning in the face of stress, fear, anger or other acute experience of emotion. It is far from impossible, in fact, and that's why you are reading this book. You know that you have the capacity and capability within you to rise above the groupthink, zombie-mode mindset which traps so many human beings for large portions of their lives.

Though the influence of society is strong, simply having a better understanding of human psychology arms you heavily to recognize situations in which you have a choice between unethical behavior in response to pressure or doing what you believe to be right or in your own best interest.

Chapter 4: Neuroplasticity and the Science Behind Forming Habits

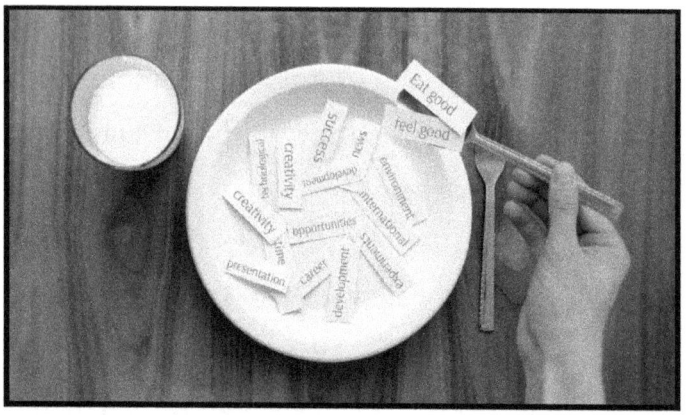

In the following chapter, we will discuss habits and the concept of neuroplasticity. If you've never heard of neuroplasticity, don't feel bad. It is actually a relatively recent area of study in neuroscience that has only developed on a large scale over the last few decades. Before we jump into how our understanding of the brain has changed in these recent years, let's take a look at some of the assumptions that scientists were working under only a few decades ago.

Like most areas of science, neuroscience has always been limited to the available technology and tools for gathering data about the brain at work. As a result, there were a lot of theories and hypotheses at work which relied on observation and the limited data that could be gathered at the time.

Scientists believed that the human brain was in a state of development until a certain point in young adulthood, at which point the brain was simply "done" changing. At this point, the brain had finished developing the essential components which make each one of us unique, like personality, disposition, demeanor, and certain thought processes. It was also believed that changing a person's habits became nearly impossible if the person waited until they were too old. It would prove to be a constant battle against the hard-wired brain connections if the person wanted to behave differently in some way.

Because of this way of thinking, people would resign themselves to the fact that after hitting a certain age, they would no longer be able to develop themselves into the vision they had for their lives. There was no real market for self-improvement, especially regarding things like mind hacking or fixing bad habits, because everyone believed that trying these things would be an impossible uphill battle. Children were pushed to learn things at a young age, and it is true that children have a much easier time of learning than adults in a lot of areas, such as language learning. It was considered essential for a child to learn the life skills and principles that he/she would need for the rest of his/her life as soon as possible. After all, they believed, "you can't teach an old dog new tricks."

But then the 21st century rolled along, and neuroscientists made some incredible discoveries.

The fact is, your brain does not stop developing in childhood or young adulthood. You have lots of neural connections in place, but it is possible to "rewire" the brain based on your conscious effort and thought processes.

So, what do I mean when I talk about neural connections and rewiring? Without getting too technical, let's explore the basics of neural connection in the brain.

Brain Structure and Habits

The basic unit of the human brain is called the neuron. The human brain contains over one hundred billion neurons. These neurons work together through their connections to other neurons to make things like emotions, feelings, decision-making, and an endless number of other brain activities possible. Each neuron is connected to up to a thousand other neurons, and these connections develop just like every other system of the body develops as we grow.

To illustrate, think of a young child who has just learned to walk her first steps. It feels unstable and scary at first. Her mother may be holding her hand as she places one foot in front of the other. But she's watched her parents and other people around her walk and has emulated other behaviors, and now she is ready to progress to walking. The brain is essentially taking its first steps in this regard, too. The brain is actively working to create a "neural network" that see her through the activity of walking. As she practices, her body adapts to this new practice; her legs get stronger, her balance improves, and her brain starts to memorize this pattern of movement and form the habit of walking until at some point, walking feels "natural" and easy.

As adults, we no longer have to consciously think about how to walk every time we want to take a step. That's because the neural connections in our brains are so strong from constant use that the roadway seeing us through this process is completely streamlined. The brain is amazing in that it not only makes connections based on what we are learning, but over time it develops a way to make the pathway faster and easier to access, like a road that was at first windy and full of holes that then gets paved, straightened out, then covered with oil for the fastest travel from point A to B possible. If you've heard the saying, "it's just like riding a bike," you already have a good idea of what I'm talking about, even if you didn't know it!

Riding a bike is generally a skill that we learn early in life, and the brain forms connections that allow us to not only learn the essential movements that go along with riding a bike, but also to strengthen those connections and improve on this skill. When you go for years without riding a bike, those connections weaken because they are not used as much. When a person first gets back on a bike after not having ridden for a decade or more, he may feel weird and off-balance at first. But the skill soon "comes back" as those old neural connections wake up and start firing again. You may not be as proficient at bike riding as you were when you were a kid, but if you decided to practice and ride your bike every day, you would find those skills coming right back to you.

These neural connections and neural networks are not just for learning basic life skills. When we adopt a behavior that later becomes what we call a "habit," we are basically describing a behavior which has become strengthened in the brain through repetition to the point that the brain automatically falls on this particular net of connections whenever they are triggered. Of course, there are many different habits, and the word sometimes intermingles with the term addiction. It is important to note that there is a big difference between a habit and an addiction. An addiction is often connected to a chemically supported impulse which the brain develops a need for in order to avoid the feelings that come with not getting that substance or whatever triggers the "feel good" chemicals that run through the body. Addiction does not always refer to substance abuse like drugs and alcohol; there are those who are addicted to social media, sex, food, etc., and these are connected to a release of those "feel good" chemicals we talked about previously when we discussed the definition of emotions. The alternative is for the brain to not release those chemicals and instead the person experiences symptoms of "withdrawal," which are never pleasant.

A habit, on the other hand, is a behavior supported by neural connections in the brain which is supported and strengthened over time. The brain has learned no other alternative to this way to accomplishing a task and has learned to default to this behavior. There are, of course, both good and bad habits. And there are numerous ways in which habit and addiction work together to form what seem like inescapable patterns of behavior. These in turn connect to certain emotional reactions, and you can see how difficult it can become when you decide you want to break free of this pattern of behavior when all of these forces are working together.

For example, take the smoker from the office who always goes outside to eat lunch, then walks over to the gazebo to have a smoke before coming back inside. There are elements of habit and addiction that overlap here to form a kind of expectation in his body that is nearly irresistible after years of conditioning this behavior and neural network of connections. When that clock on his desk hits noon, his brain kicks in to high gear, automatically associating this time of day with all the things he's conditioned it to. He will get up, he will go outside to the same picnic table. He eats his lunch, which might be one of a handful of usual lunches from his favorite restaurants nearby. Then, like clockwork, he wants a cigarette and goes to his favorite spot outside to alleviate this craving, releasing the usual chemicals into his body which signal that he has satisfied a very strong source of addiction.

In light of all this, it may seem like breaking bad habits and forming new ones is one of the most difficult things a human being could try to do, and sometimes it is. For some, it takes falling to an absolute low for them to wake up to the seriousness of their situation and how it is hurting them, emotionally, mentally, and even physically. For others, the revelation may be that they simply do not have enough good habits in their lives to get them to where they want to be emotionally, mentally, and/or physically. Each one of us is different with unique goals, habits, and life situations that must be addressed in the best way for each person. These unique facets of each human being who wants to change his/her life for the better will vary on a massive scale; but the universal truth is this—you *can* change your life, and neuroplasticity is a principle that every single person in the world can apply to do it.

Before we move on, I want to note that addiction is a serious issue, and the tips and strategies in this book may not be nearly enough to address such issues. Talk with close family and friends and consult your doctor if you feel you may need medical assistance to combat serious addictions.

Mind Hacking Exercise #2

This may seem like a lot of information, but I hope that you also find it as fascinating as I do. It's true that the human brain can be trained to throw a lot of obstacles at your feet when it comes time for you to change a bad habit or introduce a new healthy habit. But the brain is ready and willing to change according to your wishes. All it takes is a little effort and consistency on your part.

Get out that journal you've been keeping notes in (because, of course, you have been) and turn to the pages where you wrote down your progress for mind hacking exercise #1. In exercise #1, we practiced simply observing thought patterns as they came and went, noticing where our minds tended to wander and focusing on some of the things we tend to worry about throughout our days. Then, I had you write down your reflections as you watched these thoughts come and go, as well as record any emotions that are associated with specific thoughts. This may have seemed like a very simple exercise, but many people learn to run on autopilot, navigating their busy lives simply on force of habit without really thinking about what they are doing, where they are going, and whether or not that's really where they want to be. Remember, don't let anyone dictate your path for you. If there are facets of your life which feel like they've been running on autopilot, it's time to really look at these things and determine their usefulness in your life.

Review the things that you wrote down regarding exercise #1. It might be helpful to go through this exercise again if it's been a few days or even longer. See if anything has changed. What has remained the same?

Our next exercise is going to challenge you to experience the habitual brain in action. First, take out your journal and pen and write down a few things that come to mind in terms of daily mechanical habits. We're not talking about emotions or thought habits yet. What I mean when I say mechanical is something that you've trained your body to do each and every day without your conscious thought. Perhaps you walk the same way into your building at work, take a shower using the exact same sequence of movements, always hold your coffee mug with your fingers around the mug and looped through the handle instead of holding on to the handle, etc. Think of something like this for our little experiment. It doesn't have to be one of your big bad habits yet. We simply want to engage your brain and form a better understanding of neuroplasticity first. I believe having a solid foundation of understanding is important to a successful like-changing practice.

Your goal this week is going to be to adjust this mechanical habit slightly. Instead of holding your coffee mug the same way you always do, remind yourself to hold the mug a different way. Perhaps you will keep a sticky note on your monitor or something to help you remember.

The point of this exercise is to observe and record just how automatic these motor habits are and how difficult it is to adjust even the most menial element of the habit! See how you fare on the first day of this exercise. Did you need to constantly remind yourself to adjust your position? Was it easier than you expected? Did your body keep fighting to revert to the old way of doing things?

Write down your findings and continue the exercise each day for a week, or as long as you reasonably can if your days tend to vary. Keep track of your progress and especially at what point the task starts to feel easier. How long did it take for your body to start forming this new habit? Remember, every person is different, and everyone is going to form new habits at a different pace depending on the nature and intricacy of the neural network controlling that behavior. Just as with our first exercise, the real aim is to get you to simply observe these habits in action while also learning what it feels like to begin breaking a habit mentally. It may be a little frustrating at first, but keep at it.

Try to write at least a few words each day as you experiment. What is going through your mind? How does it feel when you redirect your mind to behave a different way? These emotions and mental pushback are going to come back when we dive into mind hacking more deeply to address bad habits and forming better ones, so take this time to get to know yourself and your own personal reactions. I promise, this will get easier as we go, and there are lots of things you can do to make it easier.

If you are having trouble with finding a good motor habit to address for this exercise, take a moment to look through this list and see if one of these might work for you:

- Walk in to the building where you work using a different door
- Brush your teeth starting from the opposite side of what you usually do

- Sit up in your chair with your back straight as much as possible (also a great healthy habit to continue practicing!)
- Use the opposite hand to use kitchen appliances, like punching in numbers on the microwave or setting the oven
- Sit in a different chair than usual during dinner time
- At the end of the day, kick your shoes off in a different area than usual (make sure it is ok with fellow housemates first)
- If you are usually a fast eater, try counting to five between bites during one meal a day.

Don't worry if it takes some effort just to get going with this exercise. If breaking and forming habits was easy, everyone would be in the greatest shape of their lives! Mind hacking is all about personal choice, willpower, motivation, vision, and strategy. You won't realize your goals without at least changing your mindset to one that will help you get there.

In the next chapter, we will address some of the things that often hold people back on their journeys to realizing a better way of living. We will discuss the various manifestations of anxiety and look at how overthinking and worrying affects people and their health over time. Harboring feelings of anxiety without addressing the source can really get in the way of your ability to move forward in life. We will discuss how to begin walking past these roadblocks so that you can move forward toward your new life. This step may prove more challenging than you think, but keep that vision of yourself at the end of your path, having reached your goals, to help you stay motivated.

Chapter 5: Letting Go of Worry, Overthinking, and Anxiety

We are all familiar with feelings of worry. As parents, we worry when our children go to school for the first time, when they begin hanging out with friends after school instead of coming home right away, when they start getting interesting in dating, and when they move out on their own. It is a natural tendency for people to worry about the wellbeing and safety of loved ones throughout one's life.

For many people, thought, there is a point at which worrying takes over rational thinking in situations where anxiety is heightened. They may begin to fear events or possibilities for which there is little evidence. To illustrate, think about the following scenario.

A young man named Sam is getting ready to complete three different college application essays. He has spent several weeks reviewing and researching universities around the US and their programs in his field of interest. He narrowed his choices down to three and then meticulously crafted his three essays according to each school's guidelines. He is a straight A student and has received much praise from his teachers who have also written him letters of recommendation. What in the world could he be worried about?

In today's modern world, young people are increasingly pressured to climb to the top of their competition and to stay there. Falling behind is unacceptable, especially for those who are more ambitious and capable. As the number of individuals competing for those top-tier jobs grows, the pressure to be the best is only increased further.

Pressure has a lot of influence when it comes to a person's emotional stability, especially in young people. A lot of people turn to look at each other in order to compare themselves with others. They do this to try to get a better idea of where they stand amongst the competition. Finding others "above" you in some way can feel like a crushing blow as a person scrambles to improve in their weakest areas in order to rise above.

We have to imagine that Sam has been tempted to look at how others in his own school are handling their senior years as they prepare for their own futures. Perhaps he begins to compare himself with others who have perhaps scored as good or better on entrance exams and standardized tests. Through this lens, his own personal academic weaknesses shine through and he begins to feel uneasy about whether or not these weaknesses will keep him from going as far as he wants to go in life.

It is easy for young people, especially those who are facing a milestone like this, to slip into the overthinking, constantly worrying mindset after they've begun looking around and comparing themselves to others. It is tempting to constantly use others' achievements as a measuring stick for how they are doing themselves. The minute we see others doing better, this starts to reflect on us emotionally and mentally, and this is why it can be so dangerous and self-destructive.

Sam starts to think every day about the possibility of not getting in to any of the colleges he wants to get in to, not finding the job he wants after he's graduated, finding out something is too tough for him and flunking out, etc. These thoughts, once rooted in our minds, can take hold and not let go. This is where the spiral down into anxiety can grow.

Overthinking happens when our worries and ruminations about future possibilities carry more weight than they should. Anxiety comes in to play when these outcomes tend toward being negative or what we perceive as failures on the part of ourselves or others. Anxiety tells us over and over that something has gone wrong in some way; you've embarrassed yourself, no one wants to talk to you, you'll never make the team, what if he meant something else by his remark? etc. These are just a few of the possible thoughts waiting to infiltrate our focused minds. Sam gradually loses sight of the fact that he is an amazing student with all the support of his teachers, friends and family and begins to see only the possible failures looming in his future. Hopefully, this is when someone from his support network steps in to reorient him and his mind so that he comes back to his former confident, capable mindset. But sometimes, this isn't so easy to do.

Many people move through life as fairly successful in their lives, both personal and professional, but also secretly carry very heavy loads of worrying, overthinking, and anxiety. It is not always apparent when someone is suffering like this because many people also feel the pressure to blend in and mask whenever they are struggling. No one wants to stand out as the person who "just can't handle it" and instead force themselves through experiences that actually take a great toll mentally and emotionally. I'm sure most of you are familiar with the concept of the workaholic; that student or coworker who always seems to take on way more than any one person should. These people may be feeling constant pressure to achieve more, do more, *be* more. As a result, they take on too much on a regular basis. Eventually, this person is likely to crash, most likely from fatigue. If he processes this experience as a failure, it may only fuel his need to get right back to his old routine. You can see how these thought patterns can quickly become destructive cycles.

The Role of News Media and Social Media in Perpetuating Anxiety

When occasional overthinking or worrying becomes chronic and self-destructive, it may earn the level of a genuine anxiety disorder.

There are many forms of anxiety which target different aspects of our characters and lives. One of the most influential sources of triggers for anxiety is the media.

Are you one of those people who simply cannot go a full two minutes without checking your phone for Facebook updates or text messages? If so, you are definitely not alone. Putting a discussion of social media addiction aside, let's focus in on the possible emotional ramifications of this addiction that may lead to chronic anxiety disorders. I've chosen this particular trigger because social media and phones are one of the fastest growing sources of addiction in our society today.

So, how exactly does news media and social media cultivate anxiety? Well, in the same way that Sam fell victim to comparing himself in terms of academic achievement, social media is a prime territory for people to begin overthinking themselves compared to others in a wide variety of areas. When you scroll through programs like Facebook, Twitter and Instagram, you see the faces of people and their lives. These depictions often make us feel as though we need to measure up in some way, and this need is propagated through marketing companies constantly showing us what we can achieve or earn or receive or look like if we only buy their products. This pressure combined with a constant barrage of perfect-looking, happy people is enough to fuel the anxieties of a whole generation and beyond. Eating disorders and self-esteem issues are at a record high in girls and young women. Girls as young as 9 years old begin thinking about dieting and getting thin like the models and celebrities they are constantly seeing. Men fall victim to the same pressure as well when they are constantly seeing images of the perfect male physique, which is usually accompanied by success, wealth and female companionship. These influences are not only strong but pervasive, and it is no wonder that a cycle of never feeling like you measure up can lead to chronic anxiety.

In the workplace, the constant pressure, both socially and professionally, is there to haunt adults of all ages. Those who grow older are constantly thinking of the threat of someone younger coming in to take their places. New, young professionals immediately start looking forward and measuring themselves against others who are going to be competing for the same rungs on the ladder in their careers. Everywhere you turn in life, there is another contest waiting to take hold of your mind. How can you ever break free?

Don't worry. There is hope!

Mind Hacking Exercise #3

Okay, it's time to get serious now. It's time to acknowledge those anxieties and thoughts that are holding you back and banish them forever from your life.

As always, you will need your journal or some paper or your tablet to take notes. This may be the most important exercise for you to spend time on before we move on to our step-by-step guide to rebuilding your mind. As I've stated previously, there is no sense in trying to cultivate new healthy habits before first addressing what is holding you back in life. You must clean away the clutter before filling your mind with new, positive energy and thought processes to transform your life.

Once again, you are going to have to put some mental effort into some self-examination. It's not always a pleasant experience, and we are going to be specifically targeting the nasty, negative anxieties and thoughts that circle over and over in your mind every day. You may be most acutely aware of this when you are lying in bed at night.

Many people experience sleep problems because they just can't shut off their brains at night. This is a common issue, as is reflected in the hundred of products being developed to help people go to sleep and stay asleep at night.

But very often, sleep medication is only camouflaging the issue, and there are definitive, effective steps you can take to improve your sleep quality.

The first step in this exercise is to write down the things you tend to worry about constantly throughout the day and at night before falling asleep. If you are one of those people who has a lot of trouble falling asleep, this may be your prime time for recording what you can't help worrying about. Are you thinking about money? What you have to get done at work the next day? That awkward comment you made to your spouse earlier at dinner? There are so many possibilities here. Your job is simply to take a day or two and really reflect on the things you tend to worry most about. Perhaps they are connected in some way, or perhaps it seems like a random mess of thoughts. This will become clearer as you take notes on your thought patterns.

Once you have a clearer idea of your triggers for overthinking and anxiety, we are going to engage a mental weapon in an effort to redirect the brain's wiring. You should now have a better understanding of how the brain forms networks to strengthen habits. Now it's time to start engaging this process to overcome worry and anxiety as much as possible.

First, focus in on just one of the items in your notes. It would be quite overwhelming to try to address more than one anxiety trigger at a time, so the key is to move slowly and really concentrate on the emotions that come with this particular anxiety you've chosen.

Take a moment to wrap your mind around everything this thought pattern entails. Is it connected with an emotion? Perhaps this thought makes you angry, worried, or sad. Whatever it is, try not to run away from it, but sit in it for a few minutes. What are the repercussions of this thought coming to fruition? Are these possibilities rooted in reality or is your mind running to the worst-case scenarios? Have you ever tried talking to someone you trust about this worry? What do you think would happen if you tried?

Now, once you've clearly enveloped yourself in this worry, it's time to construct your weapon.

I want you to carefully form a positive, counteracting thought that stands in direct opposition to your negative thought. If you are worried about next month's rent, conjure the scene in your mind of handing over a check to your landlord with the knowledge that you have plenty of money to cover the bill and to cover living expenses. Imagine the confidence and the happiness of fulfilling this obligation without financial concern.

Now, take a few minutes to sit in this scenario for a few minutes to wrap your mind around everything involved. How do you feel? What does the office look like? Are your children with you? Do you feel free, like a weight has been lifted from you? Form this scene in your mind with as much detail as possible. The more detail, including emotional detail, the better. This is your new weapon of choice for this thought.

The next time you feel something triggering this worry within you, redirect and interrupt the thought pattern by focusing in on your new confidence scene. Replace those old emotions with the ones you've just conjured and hold this feeling and scenario in your mind for as long as possible. It may be difficult at first, but the key is to practice consistently over time.

Record how you do during your first couple of days. Don't worry if it is quite difficult to train your mind. As we've discussed, breaking old habits and forming new ones is not something that happens overnight. Be sure to take note of when this process of thought interruption begins to feel more natural. It may take only a day or two, or it may take weeks. Each person is unique, and it's okay to be different or to take longer than somebody else.

Once you feel you have a handle on the process, move on to the next item on your list. It will probably prove too much to try to counter more than one thought process or trigger at a time. Give yourself time and take notes to keep track of how you are handling your new weapons. Perhaps you are having trouble getting over something awkward that happened at work between you and a coworker. The thought is accompanied by feelings of shame or embarrassment. You weapon may be the thought that you walk up to the person to have a conversation about the incident, and the person doesn't even remember it! All of your anxiety had been centered around this person's perception of you, so the idea that this person does not even remember what happened should go a long way to dissipate this feeling of embarrassment. Get the idea?

Take as much time as you need to in order to conquer those thoughts and feelings which may keep you from moving forward towards your new mindset. Consider recruiting friends or family you trust to help you along. They may have observed behaviors in you that you didn't even notice were happening as a stress response. Get to know yourself and your emotional pitfalls so that you can address them better. Try not to get defensive when you invite someone to talk with you and they bring up something they've noticed that they think you could fix about yourself. Changing yourself is about facing those things about you that you need to change. Don't get discouraged. It's all downhill once you've overcome all the roadblocks in your path. Your friends or family members may be able to help you dispel worries which were completely unfounded. Many times, people tend to conjure anxious or negative situations in their minds out of thin air based on others' experiences or things they are exposed to in the media. It is important to keep yourself grounded and not to fall victim to the scare tactics that often pervade not just news media but marketing campaigns. Fear can be just another way to get you to buy a product.

The next step in our journey takes us to a step-by-step guide to kickstart your journey towards a better you. The word "better" encompasses your health in every way, including mentally, physically, and emotionally. We will go through 3 chapters filled with the techniques and examples to accomplish the three things alluded to in each chapter title: First, we need to move our housecleaning mindset from the internal to the external as we identify outside influences which may be making it impossible to break free from bad habits. Second, you will create a detailed outline of what you want your life to look like. It's impossible to move toward a life goal if you don't have a clear picture of what that looks like in your life. Using similar visualization tools as introduced in this chapter, you will learn to practice seeing this reality each day to the point that it feels real in every way. Lastly, you will learn how to take one step at a time as you develop your new neural network based around self-improvement.

Chapter 6: Mind Hacking Step 1: Identify Negative Influences and Habits

The tools you learned in the last chapter are designed to help redirect your mind in the moment. As you use this tool over time, your brain will work to rewire itself until the point when you no longer are plagued by the vicious cycle of non-helpful thinking and instead defer to your place of confidence and self-worth. This is not a quick, easy process, and it will take time for the full and constant rewiring to take place. But rest assured, you've taken a huge step toward your personal goals by breaking the cycle of negative thinking.

This chapter is all about addressing the external roadblocks which may be lying in your path. Once again, you will need to break out your journal and something to write with. While the external influences holding you back may be more clear in your mind, they are often more difficult to address.

Before we get started on the writing exercise, I want you to go to a mirror somewhere inside your house. Once you are there, I want you to take a few seconds to look at yourself in the mirror. Focus in and observe the thoughts that go through your mind as you look back at yourself. How do you feel about your image? Are there things you wish to change? Are there things about your appearance you wish were different? What do you like best about your image? What have you come to terms with? Who do you most resemble? Think about how you feel in answer to these questions and any others that you can think of.

Oftentimes, when it comes to achieving our goals in life, the first and biggest obstacle is ourselves. What we think about ourselves, whether we have low or high self-esteem and self-confidence, how we present ourselves around others, or whether we tend to hide in the background. Your outside appearance and presentation is the first external obstacle you will need to address. Though feelings like confidence and self-esteem are very internal feelings, the way you present and carry yourself will have a big influence on your internal feelings. Remember the experiment where you altered your posture and observed your feelings changing? This is what I'm talking about. Even something as simple as making yourself smile will shift your attitude and confidence.

Try it now. Look at yourself and consciously adjust your posture. Keep your back straight and move your shoulders back. Use your thought interruption technique from last chapter if you start to point out things about your physical appearance that you don't like. These thoughts will never help you move forward. Instead, think of what makes you feel confident, your strengths and abilities, or your kindness and capacity for love and friendship. Whatever it is you are proud of about yourself, this is what you need to keep in mind as a weapon against self-doubt.

Hold this posture of confidence and put a smile on your face. This is where you want to be as often as possible—when you walk into a room at work or at a meeting, even walking into the grocery store. You can practice this attitude and posture wherever you go, and remember, the more you practice, the more you rewire your brain to assume the posture and attitude automatically.

Practice this in various surroundings until you feel more comfortable. As you do, get out that journal and write down how it makes you feel to adjust how you present yourself around others. What happens to your confidence in different surroundings? What is the hardest part about maintaining confidence? Anything that comes to mind as you reflect on this exercise, go ahead and write it out. Writing out your thoughts and feelings will help them become more clear. When you are ready, it is time to address possible negative influences coming from your relationships.

Addressing Negative Influences in Relationships

This may be the most difficult obstacle you have to surmount, or maybe you are lucky enough to only be surrounded by positive, uplifting people in your family and friendship circles. Each person's situation will vary greatly when it comes to how their relationships are affecting them, so, as always, we will begin with some writing to flesh out your own situation.

First, write down the people in your life with whom you are closest. The list could include lots of people or just a few. Anyone you can name whom you love and trust should be included.

Once you have your list, think about how each of these people affects you and how. If you have a partner, you may write a little bit about how this person makes you feel or supports you or respects you. Perhaps you write a little about how your mom or dad is always there when you have a problem or need to talk. This is your life and your list, so be as honest and open in your writing as possible. Part of addressing any possible negative influences in your life is acknowledging who you consider great influences in your life—these will include those who hold you up and support you in your decision to better yourself.

If you come to a person on this list and hesitate, you may have some things to think about in terms of whether the person holds you up or holds you back in some way. Now, don't get worried and think I'm asking you to cut ties with people you love. Sometimes it is as simple as having a conversation about how you need someone to support you instead of encouraging some kind of behavior. Perhaps you have a close friend who is always encouraging you to go out or drink or engage in some other activity that might not be best for you. Sit down and have a conversation with this person. Tell him or her that you are trying to conquer bad habits and develop a healthier, more successful mindset. You never know! You may just find a new partner to help you along the way. Perhaps someone on your list jumps out at you as someone who might also be interested in exploring ways to improve their lives and their mindsets. Be sure to open up and let them know how you really feel. If they are a true friend or a family member who loves and supports you, they will be willing to listen and adjust to help you. Letting yourself show a little vulnerability sends a strong signal to your loved ones that you trust them enough to open up.

Some of you will be in a more difficult position which involves a very clear negative influence in the form of a friend, family member or even your current partner whom you've been having second thoughts about. It is important to take some time to really think over how these people affect your life and what they offer you that is positive. If their negative influence over you outweighs the positive, then it may be time to seriously reconsider their presence in your life. Even family can be a negative influence if there is no shared love and respect. If you are not getting the support you need from a loved one, don't tie yourself to them just because they are part of your family. I've known many people who can attest to the fact that sometimes, it's friends who are the more positive presence in one's life, even above family. In the end, it's about who a person is inside, not necessarily whether or not you are connected by blood, which matters when it comes to living your life fully.

Whatever your situation, write down a thorough reflection next to each person's name and take note of those whom you think might be negative influences. The longer a person has been in your life, the harder it will be to think about leaving them. We as human beings love to maintain habits, and someone who is present in your life for years and years is no different from a habit that you've cultivated and grown attached to. It will be difficult to separate yourself, but if you've determined that the negative outweighs the positive, this might be essential for your life and your personal goals.

Determine who you need to talk to and set up a meeting. It is important that you do not go into the meeting with a hostile attitude. This person may not even realize the extent to which they've influenced you in a counterproductive way. The meeting should be about a conversation, not an argument. It will do you no good to get upset and start raising up your concerns as attacks. Prepare for the meeting by making a list of the things you want to address and the things you want to say.

Perhaps it's not about turning this person from your life completely but asking them to step aside while you embark on a life-changing journey.

One more thing to consider before sitting down with this person is whether or not this negative influence is something that is also causing harm to your friend or loved one. If someone you love needs help with some issue, dropping a bomb like wanting to separate from them may be too much for them to handle at the moment. If someone you love is also a victim of negative influences in their life, this might be a good time to reach out and offer to help them conquer this negative presence together. Helping others forward can be a source of strength for your own journey and determination. Consider offering to share your experiences and tools with your friend or loved one as he/she battles his/her own demons.

Addressing Negative Lifestyle Habits

Another big obstacle in most people's lives comes down to personal lifestyle habits. We form them over time, and they are often influenced by society and those around us. They start early in life, and a lot of people never acknowledge that they may be cultivating habits that are keeping them from being healthy and productive.

Our lifestyle habits encompass everything from sleep schedule and eating habits to exercise and hygiene. Everything you do on a daily basis as part of your daily life and routine makes up your lifestyle. Yep, it's time to get out that journal again.

Turn to a new page and label it "lifestyle"

For this exercise, you will need to create labels under which to list your daily habits. Some essential labels will include "eating habits," "sleeping habits," and "exercise habits." You can create as many as you like, or else simply start listing off the things that you do every single day. Perhaps you are immediately aware as we start this exercise of some things which you should be doing but are not. These may include things like cleaning the bathroom and kitchen more often or taking the stairs instead of the elevator once in a while. Again, this is your personalized list and you can choose to address whatever habits you feel are most important to address. I would highly recommend the big three which I've listed for you above, as these will have a drastic impact on your health.

Let's walk through one of these categories together. Everyone has eating habits, so let's start there.

Under eating habits, think first about how regular and consistent your current eating habits are. In other words, do you usually eat your meals around the same time or are you always eating at different times every day? Describe this as thoroughly as you can. Are you someone who tends to skip breakfast and have a big lunch? Do you enjoy having snacks between lunch and dinner? Do you tend to keep snacking late into the night? All of these different observations will be helpful. Mindless eating is one of the easiest habits to fall in to and is something that has a big effect on our bodies over time. Or perhaps you are not eating enough and not getting the nutrition your body needs to function optimally. Writing down and getting to know your eating habits is the first step to pinpointing where you've gone wrong.

Next, describe what you usually eat throughout a given week. What are your favorite foods? Do you have a sweet tooth? Is there a restaurant that you frequent several times a week because it's convenient? Do you eat a lot of fast food? How often do you cook? etc.

The reason I like to start with eating habits as an example for weeding out negative influences and habits is because everyone is somewhat familiar with what a bad eating habit is and what a good eating habit is. You know that eating ice cream late into the night is a bad eating habit, and you also know that eating a balanced dinner that gives you all the nutrients your body needs is a good eating habit.

Once you have a pretty good list going, go back through and put a star next to the biggest bad habits you would like to change.

Now, write these starred items out again on a new line and create another column across next to it. You will now decide what better habit you need to replace the old, bad habit with. That's right. It's not enough to simply know what your bad habits are; now you need to decide why it's not good for you and make a decision about what you are going to do instead.

For example, if you wrote that you tend to eat a lot of fast food throughout the week, maybe 3 or 4 meals, then you might decide you are going to focus on meal planning and pack your lunches more often to make sure you are getting healthy food. If you eat a candy bar every single day and know you need to change this habit, write down that you should bring fruit or focus on passing on the candy bar at least one day out of the week when you would usually eat it. The idea is to get to know yourself and your habits as well as the things you want to change.

You next step is to write out why you would want to adjust these habits. What benefit would changing up that sugar habit offer for your health? How would drinking more water every day affect you? Why do you want to try learning to cook healthy meals instead of ordering out all the time? In the next chapter, we will focus on visualizing your goals and making them a reality in your mind before putting in to practice the steps you need to take to make them a reality for real.

Repeat this process for all of your big categories of lifestyle habits. You might follow up the eating habits category with a list of exercise habits. It has become much easier for people to lead sedentary lives because so many jobs require us to sit for long periods of time throughout the day. You may not realize it, but doing this actually has some profound effects on your health.

Start this category similar to how you started with the eating habits. Do you have any kind of regular exercise routine? What kind of exercise do you actually enjoy? What constitutes exercise in your mind? Keep going until you have a good idea of your exercise habits. Then, write out some things that you think constitute a healthy exercise routine as part of a healthy lifestyle.

What you may find surprising in regard to exercise is that, while many people in social media brag about completing difficult 1-hour or 2-hour routines every single day, you can actually drastically improve your health by making very small changes throughout your day. For example, if you sit at a desk for long periods of time, simply making a point to get up and walk around every half hour to 1 hour will really help you counteract the adverse effects of sitting. You don't need to lift super heavy weights or run 10 miles every day to improve your health. As in an example mentioned above, simply taking the stairs instead of the elevator and taking longer walking routes to and from work will offer big rewards for your health. Introduce simple, equipment-free exercises at home like squats and lunges or sit-ups. These are all wonderful for muscle toning as well as good for your cardiovascular health.

Once you know what it is you need to improve, you can begin to create an idea in your head of what it will be like once you actually make those changes and realize your goals. In the next chapter, you will learn about the powerful impact of goal visualization.

Chapter 7: Mind Hacking Step 2: Set Your Goals and Learn to Visualize

You may remember several years back when a documentary called *The Secret* was released. The film was a huge success around the world. This film was all about something called "the law of attraction." Researchers and interviewers dove deep into history to discover how different manifestations of the same law had been passed downed through generations but had remained cleverly hidden and set aside only for the rich and powerful, or those "in the know." The Secret claimed to be releasing this secret power for everyone all over the world to use. So, what exactly is the law of attraction all about?

The law of attraction is actually a quite simple concept to understand. The challenge comes when a person begins to incorporate it into their daily lives. I will explain.

When you want something in life, you think about it. Sometimes a great deal. Think back to when you were a child and the only thing in the world you really wanted was that BB gun or that bike or that basketball. I suppose nowadays kids are more focused on electronics than old fashioned toys like these, but you get the idea!

When you're a kid, it seems like nothing else in the world would make you happier than that one thing. You include it on your birthday request list or your Christmas wish list. You do whatever you can to make it evident to your family and friends that you really want this thing. Then, lo and behold, Christmas comes around and you find your dream present under the tree. Feels good, right?

Visualize Your Goals

Let's bring this scenario up to scale and apply it to this concept of the law of attraction. The law of attraction tells us that we will naturally attract a scenario or object, or experience, or idea to us the more we concentrate our energy on it—or *visualize* it. It's really as simple as that. Now, for our purposes, we are concentrating on the visualization as part of your personal plan of action towards realizing your life goals. What this film dances around without quite saying it is that it is up to you to realize your dreams and take advantage of the opportunities that present themselves to move you closer to your goals. A million dollars isn't just going to float toward you if you dedicate three hours a day to visualizing money in your hands.

If you are interested in the unique angle on visualization and the law of attraction that The Secret proposes, it is readily available to you. This book offers a much more grounded and practical application to the concept of visualization. What we do know, through scientific research and data, is that your thoughts affect your behavior, your behavior affects your habits, and your habits determine your quality of life—that's all it comes down to. So, let's get to work.

Get out the list you prepared during the last chapter which outlined different facets of your daily lifestyle habits. We went over some possibilities having to do with eating and exercise habits, but I hope you were able to come up some personal categories for habits you would like to acknowledge and change. Remember that it's not about comparing yourself to others in your life. Their lives are not your own. You are unique, and you have something unique to offer the world. It is important to keep your goals practical and within the realm of possibility. Otherwise, you are simply leading yourself toward disappointment.

To demonstrate this process of goal setting and visualization, I'm going to introduce to you a fictional character named Laura.

Laura has been working for a tech company since she got out of college. She is a very smart woman and has done well with the company, earning the respect of both her coworkers and supervisors. After a few years, she starts to notice that the technology is rapidly changing and evolving, and she is struggling to keep up with new trainings that keep getting released. She is also noticing that positions are being taken up by new, younger professionals who are fresh out of college. They remind her of herself, though she has gotten a bit older. She begins to compare herself with these young professionals who seem completely up-to-date as far as the new technology and seem completely confident in their skills. There is a position opening up in a different department which is in a higher pay grade, and she has had her eyes on getting into this position for a long time. She talks to some friends and coworkers, and they all believe she should go for it.

She gets excited about going for this position, but as she continues to look at others in her own department, she starts to develop a habit of self-doubt. She no longer takes the lead in conversations during business meetings, and she doesn't seem to come up with as many good ideas for her group projects. Sometimes, she lies awake in bed at night wondering if she even deserves to be where she is, thought she's worked so hard to come so far.

Does anything about Laura's situation sound familiar to you? It's okay if not, I think we can still use this example as a good demonstration of how you can acknowledge your current situation, determine a better situation, then set and visualize goals.

Let's say Laura has a similar revelation one night and decides that she is ready to stop letting her negative thought cycles and bad habits get in the way of her success. What should she do first?

You're going to work through this process alongside Laura as a guide.

The first step is to create a list of concrete goals. The number of goals in your list could be as few as one or as many as fifty and beyond, whatever applies to your life. But we are only going to focus on one goal at a time, so I would suggest choosing one very important goal you can focus on and move toward right away.

Write down this goal in your journal. Now, I want you to explain in as few or as many words as you want why it is you want to achieve your chosen goal. This will help you nail down just how relevant and helpful this goal would be in your life. If the goal is superficial, you may find that your desire to achieve this goal is rooted in a place of anxiety or self-consciousness. Moving toward a goal for the sole purpose of alleviating an anxiety is something that you may want to look closely at. Wanting to lose weight because people at work have made rude comments toward you is activating a goal for the wrong reasons. This is why it is so important to not only home in on your goals but why you want to achieve them. Your goals should benefit you and your life in some positive way, not create as escape route away from the things or people who are hurting you. If you are being brought down by people at work for something like your weight, then the issue is not your weight—it is the rudeness and lack of professionalism displayed by your colleagues. If this conduct persists, it should be brought to the attention of someone who can take action to stop the behavior.

Laura's goal is to summon the courage and confidence to interview for the job of her dreams within her company. She wants to accomplish this goal because of the wealth of confidence it will give her, as well as the higher paycheck to provide for her family. She is also very happy with the work she does, and this position will afford her more control and opportunity within that field.

These are healthy reasons to want to achieve goals. They are not controlled by an adverse emotion and they benefit her not only on a material level but on an emotional and intellectual level as well. She hasn't chosen this goal to appease anyone or run away from an alternative choice out of fear. These are all signals that you can use as well to pinpoint the true motivations for your goals and whether or not they are truly long-lasting and integral sources of your future joy, confidence and fulfillment.

Once you feel solid in your choice of life goal, it is time to visualize yourself in a scenario in which you have been successful. We've practiced visualization in previous chapters as a way to combat negative thought cycles. Bring your skills for visualization into play here once again and turn your focus on this future joy in accomplishment. Let's walk through Laura's visualization for having accomplished her highest goal.

Laura sees herself waking up in the morning feeling refreshed and ready for her big interview. She gets dressed and feel confident as she heads to work. Maybe she's playing her favorite radio station or listening to her favorite podcast.

When she walks in the front door of her office building, she is greeted by the front desk and she responds with a warm smile. When it is time for her interview, her boss calls her into his office and invites her to sit down. She maintains great posture and a pleasant, confident demeanor as she answers each question pertaining to her readiness for this new position. She answers each question naturally and even throws in some humor here and there, which makes her boss laugh.

At the conclusion of the interview, her boss stands up and offers his hand for a handshake with a warm smile. Laura returns the gesture and thanks him for his time, then exits the office.

A few days later, Laura receives a phone call shortly after she's left work for the day. She is offered the job and will start the following Monday. Her face is beaming as she thanks her boss for the call and hangs up. Maybe she does a little victory dance outside her building and plays her favorite song from her phone on the way home.

The following Monday, she is escorted into her new office. She will get to decorate the area however she wants, and her office offers a new, wonderful view of the city from higher up in the building. Her boss comes in shortly to introduce her to some of her new coworkers as well as go over her initial job duties. Laura feels confident as she listens and nods her head as her boss goes over each step to the day.

Laura can close her eyes and visualize the moment when her boss comes in and congratulates her on her achievement. She imagines the weather outside, what she is wearing, how she's done her hair, and how others around her interact with her. Everyone is responding to her positive energy, and she seems to light up the rooms whenever she walks by.

This is an example of a visualization that turns a person's goal into reality within the mind. This is exactly what I'd like you to do with your own goal, no matter what it may be. It may be completely different from Laura's visualization, and that's fine. The reason I went through an example is to illustrate how much detail to give your own visualization. Really make it real in your mind with as much as you can add to the scenario.

Visualize New Habits Through Mindfulness Practice

Now it's time to revisit your top goal as you've written about it in your journal. You are going to create a new column next to where you've described this goal and why you want to achieve it. You task now is to list some new habits that you can work on to bring you closer to you chosen goal. Once again, we will use Laura and her goal as an example.

Laura has thoroughly imagined herself living out her dream of summoning confidence and courage to perform perfectly in a job interview and land the position she wants with in her company. Now that she can see herself clearly in this place, it is time to create a plan of action to start making this dream a reality.

What are some of the tools Laura will need in order to confidently conquer her self-doubt? Some of her healthy habits she'd like to pursue might include:

- Positive thinking and mindfulness on a day-to-day basis

- Daily mantra for confidence, sticky note reminders throughout the day
- Have one positive conversation with a coworker and/or supervisor each day
- Think of something to improve in terms of workflow and share with others
- Be proactive in meetings and brainstorm ideas to present for projects beforehand
- Practice being prepared for anything that comes up throughout the day
- Make time to work out every day or every other day
- Focus on a healthy lunch at least 4 out of 5 days of the work week
- Set aside time to do some extra research on the technology you're not confident with
- Ask questions and don't feel bad about having to do so

Laura's list could go on to include many more possible actions to take, but this is a good starting point. Notice that each of these courses of action move her closer to being prepared mentally and emotionally to complete her own challenge. She may focus on just one or two things at a time or try to tackle several at once as she incorporates them into her day. It is important that you take some time to really think through your action plan to determine which habits would be the best focus for you and your goal.

Create a similar list of items for your to focus on that will bring you closer to realizing your goal. In the next chapter, we will move on to how best to teach your brain to adopt these new habits, but one of the most important things you can adopt into your life is something called mindfulness, and we will discuss this concept briefly here before moving on.

Mindfulness is mentioned all over the place in conjunction with various spiritual practices, religions, and products. Mindfulness is actually a very simple concept that you can begin using in your daily life right away. It's all about paying attention to what is happening to you *here* and *now*. Too often people are worrying about or thinking about what's going to happen tomorrow, next week, or next year. They worry about what happened to them yesterday or last week and stress about what people think of them, etc. When we let our minds focus on things outside of the here and now, we really miss out on a great workout for our brains. Mindfulness has actually been shown to help the brain repair itself while clearing out all the clutter that is thrown at throughout the day in the form of information overload. We will get into mindfulness meditation in a few chapters, but what I want you to do is simply try to be more mindful of where you are in the present moment.

The practice is similar to how we created mental weapons against negative thought patterns. Instead of interrupting bad thoughts and introducing better thoughts, try interrupting your wandering mind and bringing it back to the present. The impact on your brain will be just as powerful as you are breaking down and forming a new habit, just like in our other exercises. It may seem simple, but practice over time leads to a huge difference in cognitive function, memory and overall mental acuity. You will also feel happier more often as you let the burden of overthinking lift from your shoulders. You will start to realize that a lot of what we all worry about on a day-to-day basis is ultimately useless and irrelevant. These thoughts can also become emotionally harmful, as we've demonstrated in past chapters. Protect your mind and fortify it for the tasks to come, because we are just getting started!

In the next chapter, we will zero in on a few of your healthy habit goals and learn how to slowly introduce the new habit into your life step-by-step. It may be helpful to see if you can recruit a friend to adopt a new healthy habit with you for support. You will find that once you've gotten the hang of how to slowly introduce change into your life, you will want to keep going as you begin to see the benefits of your new healthier lifestyle reflected in your body and mind.

Chapter 8: Mind Hacking Step 3: One Step at a Time

It's time to really get in to the nitty-gritty of changing those bad habits into good habits.

You should have some idea of the things you know you need to change in order to move closer to your goal. The idea here is to introduce one small change at a time and give yourself ample time for your brain to rewire itself according to that new habit. You want to concentrate on one small step that you will address every single day for three weeks. That's right—three weeks. Once you've developed this new habit, you will begin to build upon it by introducing a new habit for the next three weeks, and so on. You can choose to adopt just one habit to begin with, or you may feel confident that you can adopt a few different ones at once. It may be easiest to combine physical and mental habits so that you don't overwhelm yourself with a bunch of mechanical habits all at once. Remember our exercise with mechanical habits? It is difficult to retrain the brain to do something different from what it's been used to. That's why combining some kind of mental reinforcement with a physical habit is a good combination rather than trying to introduce several mental exercises into your daily life right away.

Once again, we can use Laura to demonstrate this process in action.

Taking her cues from the list we created in the last chapter, let's say Laura wants to focus on mindfulness and having one positive interaction with a coworker or supervisor every single day. The first thing Laura needs to do is prepare herself for the first day. She may already be familiar with mindfulness but watching some videos or reading some further instruction may be helpful. Her mindfulness practice will center around a 20-minute morning routine that will focus on her day, her goals, and her mindset. She may write out her own mantra to help her get started, and it might sound something like, "I will hold my head up when I enter work and keep a positive attitude throughout the day. Nothing can stop me from reaching my goals. I am strong. I am confident." Once she has prepared her mind, she is ready to tackler her day.

When she arrives at work, she may notice a supervisor walking past whom she hasn't had a real conversation with in a long time. Laura feels a little nervous about approaching her, but then conjures the courage to walk up to her.

She greets her supervisor and the greeting is returned warmly. Laura may find that the supervisor already knows a great deal about her and her ambitions. They talk for a while, and Laura maintains her positive attitude as she's promised herself she would do. When the conversation is over, she feels a wave of satisfaction and relief. She knows she is capable of conquering fears, and she sees that the key is to take small steps at a time.

So what about your own goals? It's time to choose one of your habits you want to adopt and begin making this change every day. Remember to choose according to what you feel comfortable taking on at first. You can add to your habits later. Perhaps you are determined to quit smoking and have decided to replace this behavior with something healthier, like doing some exercise during the times when you usually take a smoke break at work.

This is a good technique if your personal situation dictates that you need to replace one bad habit with one good habit. Sometimes simply focusing on breaking a bad habit is overwhelming without something to replace the behavior. Your brain needs something to fill the void and begin rewiring its response to the trigger. Some other examples might include: swapping water for pop, walking or biking to work instead of taking bus or driving, taking stairs instead of the elevator, packing a lunch a couple days a week instead of getting the same fast food you always get, greeting your boss with eye contact and a smile instead of a down-turned head, joining in on a conversation instead of avoiding people during lunch time, and practicing a few minutes of meditation in the morning instead of hitting the snooze button several times.

One method that is commonly used to help keep a person motivated while changing habits is to keep a calendar somewhere prominent in your home. Alternatively, you can use an app on your phone. The idea is that you have somewhere to mark each day that you successfully complete your new habit or avoid your bad one. A person will take anywhere from a couple of weeks to several months to break a habit or form a new habit, so don't worry if it is taking a little longer than you expected. It doesn't mean you are less capable; each person simply has a unique pathway toward learning.

Once you have your plan in place, prioritize your one simple change each day. No step forward is too small. If you find that exercising every single day after work is not something you can do, redesign your daily habit to something that you feel confident you can accomplish each day. Focus on just 15 to 20 minutes of exercise a day, at whatever time is convenient. It may be difficult to stick to the same exact time every day, so personalize your task so that you are giving yourself the best possible chance of success. Each day, make a point to acknowledge your accomplishment by checking off that day on the calendar or in your app. This is important, as it gives us a feeling of accomplishment and offers a big of a boost to keep the routine going.

Once you have a bit of a streak going, you will find it easier to continue along with the momentum you've got going. Human beings are hardwired to be somewhat more loss averse in comparison to desiring a gain of some kind.

For example, as data has shown in various research studies, people tend to fear losing 5 dollars much more strongly than they desire to go after an additional 5 dollars. That idea of loss simply carries more weight. This principle will help you as you begin to see the checkmarks adding up across your calendar. At some point, you will start to feel more and more strongly about not breaking the streak you have going. It feels good to know that you are capable of this daily change. And within no time, you will have formed a new healthy daily habit.

Positive Reinforcement

You are probably familiar with the concept of positive reinforcement, even if you can't perfectly explain it. Ever heard of Pavlov's dog?

Pavlov's dog refers to an experiment that was conducted by Ivan Pavlov where he successfully trained a dog to associate the smell of food, which made the dog salivate, with other stimuli which would not usually make a dog salivate, like the sound of a bell. Before long, the dog would salivate at the sound of a bell, even if there was no actual food around. This is one of the most famous examples of what is now known as associative learning.

The dog goes through a series of conditioning that ultimately leads it to associate salivation with a neutral sound.

When we introduce positive reinforcement into the picture, there is another dimension to the learning process that involves a reward system. Positive reinforcement is simply when a specific behavior is rewarded with something desirable. For example, you might tell a child that if he behaves while he's out with mom running errands, then they will stop and get some ice cream on the way home. This reward, and anticipation of a reward, motivates the child to behave while he is out doing boring things with mom. Positive reinforcement works to support learning only when the reward system is consistent. Another example of positive reinforcement is the little shiny stickers teachers might put on a child's homework assignment to tell the child he did a good job. Each time the child turns in a homework assignment, the assignment is returned to him with one of these stickers. The sticker is a symbol that says the teacher is proud of the work the child has done. Children like to impress their parents and teachers, so this often serves as a powerful source of positive reinforcement.

This idea can be applied to your own life as you discipline yourself to follow your daily habit regimen. It may be enough for you to look way down the road and be able to see yourself having successfully formed your new habit, but some of you will find that you need an extra boost of motivation every now and then to keep going. It makes sense that this happens.

Sometimes, the reward for realizing a positive new habit doesn't come for a long time, even if you know it will be very beneficial over the long term. Human beings tend to be drawn toward instant gratification, and this is encouraged by services that deliver exactly what you want either immediately or within a short amount of time. Just about anything you may want is available online and can be shipped to you within a single day, if you choose to pay for it. A hundred different delivery services will go get the food you're craving and deliver it to you within a half hour. If we want to see a new movie or TV show, all we have to do is get online to our favorite streaming services.

Our society has evolved to encourage instant gratification, and, as a result, a lot of us tend to struggle with the idea of working for the long-term and patience in getting what we want. You've seen the ads and commercials depicting a woman who fails on her diet because she is too tempted by junk food or sugar. She may really want to lose weight, but the lure of sugar is too much for her to overcome.

Because of this phenomenon, it may become necessary to introduce a system of positive reinforcement as you are in the fragile beginning stages of forming new habits. This means that you will reward yourself for the fact that you've kept up with your daily goals periodically throughout your journey. It could be every week or every couple of weeks, maybe something a little bigger once a month. As always, getting out that journal and writing out some ideas will help you come up with a plan that will motivate you.

Think of something that you really enjoy and that would be a powerful motivator for you to accomplish your goals. Perhaps once a week you will treat yourself to some of your favorite ice cream or let yourself buy some clothing or a pair of shoes. Maybe if you can complete two weeks of following your daily tasks, you treat yourself to a pair of tickets to a sports game you'd really like to see. Whatever motivates you and excites you, use it as a way to keep yourself going during those times when the daily task feels more like a chore.

I promise you will feel this on occasion, and it is perfectly normal.

Another tool that I come back to again and again is simply setting aside time every single day to write about what you are experiencing. Many times, our feelings are not very clear, and taking time to write and flesh out the origins of our emotions can go a long way in helping us address the source of those emotions.

If you are feeling tired, or perhaps you miss a day of your habit-forming task, write down in your journal why you think you felt that way or lapsed on your daily task. Is there something else in your life that is bothering you or causing anxiety? Perhaps there is a distraction that needs to be dealt with. This is why I spend so much time emphasizing the importance of rooting out negative influences, both internal and external, before embarking on this habit-forming journey. But sometimes there will be things outside your control that you have to deal with before moving forward, and that's okay. Take some time to address what is holding you back, just like we focused on doing in previous chapters, then take some time to regroup before trying again. Remember, failure is an inevitable part of success, and there will be obstacles to overcome as you work toward your goals. Give yourself as much support as possible by doing your best to remove sources of anxiety and stress from your life before introducing new habits.

As you regroup, write about how you are feeling. Perhaps you don't know exactly what led you off track. Go back as far as you can as you recall your progress to pinpoint exactly where things went wrong.

If you are simply dealing with a lack of motivation, try writing about how far you've come and what it will mean to you when you achieve your goals. Return to that visualization of yourself having accomplished your goals as often as possible and as often as you need; that's why it's there. If you can keep this visualization in your sights at all times, especially when things start getting difficult, you will be giving yourself extra support and motivation to get through the rough spots.

Another source of support may be available to you through trusted family, friends, or romantic partner. Set up a time to meet and talk over coffee or something and explain what you are trying to do. Be open and let your feelings out about whatever doubts you have about whether or not you will accomplish your goals. Feeling doubt is not a failure on your part, it is natural. Also, feeling self-doubt is a strong sign that what you are doing is important to you, because you are invested in the outcome and realize what is at stake. Keep reminding yourself of the reward to come.

When you combine all of these strategies for support, you will find that you have a much stronger foundation from which to move forward. Keep journaling, ask for help or just talk with friends and family, integrate a system of periodic positive reinforcement, and keep your personal vision handy at all times to combat thought of self-doubt that come up. People are usually stronger than they think in the moment. Arm yourself to be as strong and as likely to succeed as possible.

Negative Reinforcement

Negative reinforcement is often misunderstood as the equivalent to punishment, but this is incorrect.

Negative reinforcement refers to the taking away of a stimulus, the act of which encourages a specific behavior. A good example is the little beeping alarm that goes off in your car when you don't have your seatbelt. This is a form of negative reinforcement because you are encouraged to buckle your seatbelt because you want the annoying beeping sound to go away. It is the counterpart of positive reinforcement in that it also encourages a certain behavior, but instead of adding a positive reinforcer, you are simply getting rid of a negatively impactful stimulus. Punishment, on the other hand, is the use of a negative consequence of a certain behavior. You steal from a store, you go to jail as punishment. This is very different from negative reinforcement.

Negative reinforcement can also work for you in terms of strengthening your resolve to keep going with your daily habit-forming tasks. To use the example of quitting smoking again, you will very quickly begin to see the physical benefits of taking away the cigarette smoking. You are rewarded, through negative reinforcement with an improvement in overall wellbeing, energy, food tasting better, breathing easier, etc. The list of benefits is extensive in this example. If the idea of negative reinforcement applies in your personal situation, take some time each day or every couple of days to take note of the benefits you are receiving as you remove that negative thing from your life. Write regularly and keep going back to those pages each time you feel challenged to keep going.

Moving Forward

You should be proud of yourself for beginning this life-changing journey! You now have the essential tools you will need to be successful. Revisit these chapters as you make progress and remember to share your success with others for additional support. There is nothing better than receiving a hug or acknowledgement of your work from those you love and trust.

In chapters 9 and 10, you will learn how to introduce additional tools into your new life practice, including meditation and a list of 10 daily practices to strengthen self-confidence.

Chapter 9: Meditation Techniques

Meditation has become one of the most popular self-help techniques to ever reach the Western world. There are countless books available offering extensive history and instruction for newcomers who would like to learn what meditation is all about. In this chapter, I will give you a general introduction to the practice of meditation, why it is useful, and how you can begin introducing it into your daily life as a way to strengthen your goal of forming new habits.

Why Meditation?

The practice of meditation has been around for a very, very, very long time. Until fairly recently, meditation was usually tied in most people's mind to a certain religion of spiritual practice. There was always a whole history and spiritual tradition behind the actual practice. The most familiar example of this is Buddhism.

Buddhism is nearly synonymous with meditation in a lot of people's minds. The story of the "first" Buddha and his enlightenment is taught as the first stepping stone toward realizing each person's inner "Buddha nature" within themselves. Consistent meditation practice over a period of time is how people get to this idea of enlightenment, though the syntax for believers assures that there is no journey toward, but simply a realization.

Meditation became mainstream in the Western world, and soon there were lots of different sects and groups of people trying to popularize their own group's meditation "style." One of the most well known might be the Transcendental Meditation movement, popularized by such figures as David Lynch and Gwyneth Paltrow.

The fact is, meditation is available for anyone and everyone to learn and practice. Some people will latch on to certain styles and spiritual practices and belief systems associated with it, but this is not necessary. You do not have to be religious or adhere to any set of spiritual beliefs to benefit from a daily mediations practice, and that's because the effects of meditation are scientifically documented and well-researched.

Something very profound happens in the brain during a deep meditation. During meditation, the practitioner is essentially cleansing and repairing the brain similar to the process that occurs during a deep sleep. Similar to the benefits of sleep, meditation encourages the organization and solidification of information necessary for long-term learning and muscle memory.

For example, perhaps you yourself or someone you knew in childhood trained in a musical instrument, either for school or personal enjoyment. You would study and practice with a teacher for an hour or two, then at the end you kind of felt overwhelmed with information and things you needed to practice.

You may have found that after a good night's rest, you would wake up and feel refreshed with a better organization of information in your mind.

The next time you sat down to play, it seemed easier and more clear to get out the notes and play a piece that you may have been struggling with just a day or two previously.

This is the brain hard at work for you as it works to make the pathways and neural connections as streamlined as possible. Meditation has the same effect when it is practiced consistently over time. This is why meditation would be an excellent tool for you to introduce into your daily life as your brain works to form new habits. It doesn't have to be a long meditation; just 20 minutes a day will go a long way. We talked about mindfulness earlier in this book, and you can easily adapt this principle of mindfulness into a concentrated 20-minute daily meditation through a variety of techniques.

Mindfulness Meditation

When we discussed mindfulness, you learned about the practice of reining in your thoughts to focus on the "here and now" as opposed to all of the other things fighting for space in your mind. You practiced focusing your thoughts and focusing on what is going on right before you instead of something that happened last week or your to-do list for the next day or the next week. It is not easy to adopt this practice at first, especially for a complete beginner. But there is nothing better for your brain as you begin training it to change and form better, healthier habits.

In this section, I will walk you through a basic meditation that you can begin implementing into your life immediately. You don't need to pay for special instruction or study a thousand years of history to understand and experience the power of meditation.

First, you will need to find a space inside your home that is comfortable and free of distraction. This may be more challenging for those of you who live in apartments with roommates or a similar situation but try to find a space either inside your home or, if the weather permits, someplace outside that is quiet. I wouldn't suggest going out to a public park or a similar setting for your first few sessions as the background noise might be quite distracting.

Once you've found a quiet spot, assume a comfortable position. You don't have to twist your body or legs into complex positions like you've seen people do in movies or advertisements. Sit however is comfortable for you. If you decide you want to get more serious about meditation in the future, there is a ton of information and history waiting for you to discover in order to learn more complex techniques, such as Zazen according the Zen Buddhist tradition. For now, your goal is simply to remove any stress and tension from your body so that you can completely focus on what's going on inside your mind.

To begin releasing this tension, we will start with something called a "body scan." Bringing your awareness to the present moment entails that you are also as aware as possible of your own body in space. It may sound weird, but we actually go through most of our day-to-day lives without consciously feeling many areas of our bodies. Our brains simply choose to bypass this awareness in favor of a task at work or whatever else our daily routines demand of us.

To start with a body scan, first bring your attention to your toes and feet. Move them slowly to help you with this, then begin taking slow, deep breaths. Thinking about your feet, slowly begin to relax and release as much tension as possible from your feet. It may help to first tense the muscles, then slowly relax the muscles. Visualize the tension floating away on the air.

Next, move your focus to your ankles, your calves, the rest of your legs. Gradually focus on each part and slowly relax as far as possible. You may find that the best position for this exercise is lying flat on your back to help you relax.

At the halfway point, take a few minutes to focus on your breathing. Don't worry if you have trouble focusing. Your mind is going to want to wander on different things, and that's okay. Don't try to force your mind blank. Instead, when you realize your thoughts are wandering away, simply restore your focus on the task at hand whenever you realize what's happening.

Try not to get frustrated with yourself if it is hard to focus at first. Meditation is not something that comes natural to everyone, and as we all continue through our lives in an information overload society, it makes sense that the brain has trouble calming down. Simply redirect your thoughts and come back to the present moment. Focusing on your breathing is one of the simplest tools you can use for practicing mindfulness and throughout your dedicated meditation sessions.

Now, move the focus up to your torso. Put your hand just under your ribcage and feel the breaths you are taking as they come from deep within, then release. Do this several times, feeling the breath as you inhale, then slowly releasing. As you exhale, focus on relaxing and releasing as much tension as possible.

Finally, move the focus throughout your arms, then move up to your head. The last point is the very top of your head. Hold this focus for several breaths, then release.

Congratulations! You've just completed your first meditation. If you are brand new to the practice, I would suggest trying to fit in just a few minutes of meditation every day while you practice your new habits, but don't overwhelm yourself by trying to adopt a 20-minute meditation routine right away. If it works for your set of habits, it might be beneficial to try and schedule your daily habitual task alongside a meditation session. Keep track of how you feel and write down notes to keep track. How do you feel directly following a meditation session? Is it something you feel you can benefit from on a daily basis? Keep in mind that you may be able to find a nearby group or meetup that practices meditation. Having a mentor and teacher to help you develop your practice is the ideal way to get better and maintain a daily meditation practice. If the group meets weekly for meditation, consider incorporating this into your habit-forming task list. As you use meditation to clear you mind and focus, your dedication and ability to maintain your daily habit-forming task streak will become stronger and stronger.

Guided Meditation 2: Mantra

A mantra is simply a script that you use as the focus of your meditation session. There are thousands of different mantras and scripts, and you can also choose to create your own. You may have heard of the practice called "loving-kindness" meditation. This meditation entails focusing on an outward radiation of love and kindness towards the world through scripts that reinforce this feeling of love and oneness. A simple Google search will provide you with tons of examples of scripts to use for this mantra meditation. We've actually mentioned one already in the chapter where we used Laura and her daily task of reciting words of self-confidence each morning. Again, you can choose from a list of others' scripts or create your own. The idea is to choose something that is central to what you are trying to cultivate in your heart and within your mind.

Find your comfortable spot in the house or outside if the weather is nice and there are not too many distractions. Relax your body and settle in to a comfortable position. Take a few deep breaths to help you relax. Observe your thoughts as they wander but don't force your mind to go blank or push thoughts away. Let the thoughts happen, then gently redirect back to your breath and present space. Begin by visualizing yourself in front of you, the stresses of your day weighing heavily on your back. Visualize these as stress that is captured inside a helium balloon. Focus in on each one. Maybe you had a difficult customer you had to deal with that day, or you had an argument with a coworker. Maybe you feel nervous about an upcoming bill, or you have an obligation coming up that you are not looking forward to. Whatever the source of stress, you are going to gently let go of each one through a visualization exercise.

Concentrate on one stress at a time, then visualize your hand on the string of the balloon as you let go and the balloon begins to rise. Watch in your mind's eye as the stress begins to float away from you, higher and higher into the sky. As you watch, the balloon gets smaller and smaller until you can barely see it, then it disappears. Do this for each of the triggers for stress that your mind keeps moving toward. After you've released the last balloon, take a deep breath alongside the picture of yourself in your mind. Now, recite your mantra.

Either write down or memorize your chosen mantra and begin saying the words slowly aloud. Really focus in on the words and what they mean to you. Let them build you up like you are putting on a suit of armor. Feel the conviction in your voice and internalize the words. Reinforce the mantra by repeating aloud the words, then silently contemplate their meaning. Continue to visualize yourself embodying the attitude of the words in your mantra. If you are concentrating on a loving-kindness mantra, imagine yourself surrounded by others with whom you are sharing warm smiles and hugs. If you are focusing on your own inner strength, imagine yourself performing a daily ritual with the conviction of strength and confidence you want to maintain. Whatever the words of your mantra lead you toward, imagine yourself in a position of strength in conjunction with those words. The mind will internalize this feeling of empowerment, and you will be able to call on this energy when you need it.

Basics of Zazen

Some meditation practices call for the practitioner to try to focus hard on a mantra or other point of focus. Other meditations want you to try to erase your thoughts and quiet your mind through conscious effort.

Zazen, or "seated meditation," is an ancient meditation practice that is tied in with Zen Buddhist philosophy and teaching. I want to briefly introduce this tradition, as many people find it to be one of the most transformational and wonderful experiences of their lives to utilize this meditation daily.

Formal Zen meditation calls for a certain posture and body position, but you don't have to try tackling all of the details at once. This meditation is all about watching your thoughts as they move without hindering their movement. It is different in that you are not trying to exercise control, but instead go with the flow according to the principle of impermanence. A thought will land on a certain thing, it may linger a bit, then is moves along. There is no standing in place or freezing the mind on one subject because that is not the natural order of things.

To experience what I mean, go ahead and assume your comfortable position and begin simply watching your thought patterns. We've touched a bit on this during our focus on mindfulness, but instead of trying to reign in your thoughts and redirecting, simply try not to dwell on any one thought for longer than a few moments. Watch the movement as the thoughts come and go and where they tend to linger a little before moving on. Do you keep coming back to some stressful recent experience? Perhaps your mind wanders into territory where you begin to feel anxious about possibilities. As soon as this thought enters your mind, try to gently brush it along as you move on to a different thought. The idea is to not get "stuck" on any one thought and to watch the continuous movement through your mind.

In "proper" Zazen practice, the practitioner will sit up straight, perhaps with the help of a small cushion, with his head gently relaxed and slightly tilted forward. The eyes are slightly open and are directed at the ground just in front of him. Arms rest in his lap with one hand resting in the other, the thumbs forming a small "O" in the center. The added discipline for forma meditation adds a whole new dimension to the experience and should be reserved for when you become more practiced. The most important part is getting started and getting your mind used to the jolt of redirection and focus.

Whatever style you choose to pursue, I highly recommend that you choose one to incorporate into your daily routine. Meditation can be a powerful and rejuvenating way to start your day if you can set aside just 20 minutes before heading out the door for work or whatever daily tasks you have planned. Even if you are simply setting aside a few minutes to practice mindfulness, bringing your attention to the present, this is going to form a powerful source of strength and motivation for you as you build better habits.

Give yourself time to master techniques such as the ones mentioned in this chapter. It is a skill, just as learning to rewire your brain is. Trust your capacity and ability to form new habits through simple repetition and conviction.

Believe in what you are doing and where it will lead you.

In our final chapter, we will go through a list of 10 useful daily practices that you can incorporate into your routine to support self-confidence. A breakdown in self-confidence can be a big blow to your ability to keep moving forward. Reinforce your determination with these tips to keep you going, even on those challenging days when you need extra motivation. As always, never lose sight of your personal goal visualization, standing in that victory pose with a giant smile on your face!

Chapter 10: Moving Forward: 10 Daily Practices to Strengthen Self-Confidence

Building self-confidence is essential to maintaining the attitude and the energy to persevere in your journey towards a better life through better habits. As you check off those days and watch the chain grow across your calendar, there are several things you can do to help you stay motivated. Integrating a daily meditation practice is an excellent way to help your brain as it is challenged to rewire itself towards a better state.

In addition, I'd like to offer the following 10 daily practices for building self-confidence. All of them are simple and easy-to-use tools to help you stay focused and your mind to stay on track.

Daily Practice #1: Compliment a Stranger

Your first daily practice tip is all about going out of your way to compliment a stranger. If you are a shy person, then this may be a little more challenging to you than to someone who is naturally outgoing, but the benefits of accomplishing this daily task will be powerful for anyone who undertakes it.

You don't have to overthink this practice. Simply go about your day but be conscious of the people around you. Not only will you be more mindful of how you can help those around you, but this also allows you to incorporate a valuable practice of mindfulness into your day. For example, say you are at the grocery store and you are waiting in line with several other people. Instead of blankly staring at the gum, think of something nice to say to the person either in front of you or behind you. It may be something as simple as complimenting a piece of clothing or jewelry. Or perhaps you want to tell them they have a beautiful smile. I promise you that just a simple gesture like this one can light up a person's whole day.

Think of a time when this happened to you. It meant a lot, didn't it? It seems like such a simple gesture, but there is a profound power in human connection, even in something as simple as a compliment. You may find that you love the feeling so much, you start practicing complimenting strangers automatically. The key is to stay mindful of what's going on around you. Everything in that grocery store or in your workplace or school has a special set of circumstances affecting their lives. Many of them are struggling with something challenging, so taking the time to say something nice to someone can have a much larger impact than even you can imagine in the moment.

Daily Practice #2: Lift Weight

Now, don't panic! Let's look a little deeper into this tip.

We talked a little bit about getting more exercise every day and how it isn't necessary to do a monster workout every single day to improve your health. This is still the truth, and there are lots of options for you to choose from that will only take a few minutes out of your day. It will be well worth it.

Consider investing in a pair of dumbbells. They are not expensive, and you can tailor the weight according to what you are comfortable with. You can get hand weights as light as 3 lbs. all the way up to 40 lbs. and higher. Lifting weights for just a few minutes a day is going to benefit you in many different ways.

First of all, it's obviously good for muscle toning, and you also get the benefits of those feel good chemicals we discussed early on in the book in relation to emotions. The chemicals released into your body as a result of exercise is a natural high that boosts your mood, as well as your self-esteem. Lifting weight gives you a feeling of accomplishment that is immediate and long-lasting. You will continue to feel good for a long time after challenging yourself with some weights.

And you don't necessarily need hand weights to get the benefits of this type of exercise. There are several different exercises you can do which challenge you using your own body weight. Some examples include: pushups, plank, burpees, lunges, and squats. You will find plenty of resources online to show you how to properly execute each of these exercises. They don't take up much room, and a simple 10 to 20-minute workout is all you need to focus on to begin adding this daily practice into your routine.

The confidence streams naturally from having accomplished a weight-lifting feat. You may find that you love the feeling so much, you want to gradually add weight to add to the challenge, and that's great! Be careful not to challenge yourself too far in the beginning, especially if these exercises are new to you. Remember, try to do a little bit even if you don't feel like it. The simple action of beginning the workout will give your energy a boost even if you were convinced you had nothing in you that day.

Daily Practice #3: Watch a Motivational or Inspiring Video/Presentation

This is one of my favorite practices. It requires very little effort on your part and in return, you receive a boost in mood as well as self-confidence.

We human beings favor storytelling when it comes to passing on lessons, from fairy tales and fables teaching us valuable ethical behavior to historical study that tells us all about our triumphs and failures as a species. Taking time each day for a motivational video or even just to sit down and read some motivational quotes by famous speakers will help raise your self-confidence and also get you excited about moving forward to new challenges. There is nothing better than a good speech or talk. Head to YouTube or Google a favorite figure to see if you can find just what you are looking for. There are so many things to choose from. TED Talks are plentiful and easy to find on YouTube. There are TED Talks that are free to view on YouTube that cover an endless list of topics from a variety of speakers from all over the world. Choose a topic that you are personally interested, or else find a talk which directly addresses topics like confidence, conquering your fears, and forming good habits. There is a great deal for you to discover, and it just takes a few minutes to reap the reward of a good old-fashioned pep talk.

This daily practice is especially helpful on days when you feel less motivated than usual. You may feel like you have less energy and are just not on top of things the way you usually are. I promise, after a few minutes of listening to a motivational speech from a respected figure, you will feel renewed and imbued with a fresh surge of energy and positivity. The feeling is contagious, and when someone speaks from the heart with the intent of spreading it to listeners, it is impossible not to pick up on that energy. Use it to keep you going throughout the rest of your day.

Daily Practice #4: Learn Something New

The brain needs regular exercise just like the body does. This daily practice is all about using your brain power to improve your cognitive function, memory, and learning. It may sound like a lot to take on, certainly for a daily practice, so I'm going to suggest a plan of action to get you started with this.

The idea is to engage your brain in a new skill. It doesn't have to be super challenging or overwhelming. Unless you are really motivated and interested in doing so, I wouldn't suggest buying a college textbook on chemistry, for example, and trying to teach yourself all of the lessons. Learning something new can be as simple as reading an article on a new piece of technology in the newspaper. Perhaps you have a special interest in something like planes, or historical war tactics, or psychology, or flight patterns of birds. Whatever sounds interesting to you, look up a short article that will teach you something new about that topic. I would suggest making sure to check the sources for the articles you read, as some are going to be of better quality and more researched than others. This new knowledge will make you feel stimulated and will wake up your brain, helping you to focus on other tasks. It is too easy nowadays to simply let your mind melt while you take in hundreds of images and mindless text from social media and other apps on your phone. Be proactive with your brain's development and feed it something with a little more meat on it! You brain will thank you. It feels a lot like finishing a meditation session. Concentrating on a scholarly text helps to flush away needless background information as you bring all of your faculties to focus on the words you are reading.

Alternatively, perhaps there is a skill you've always wanted to learn but have never had time for. Set aside some time each day to practice learning a new skill that you will enjoy. It shouldn't feel like a big chore, because if it does, you will find it very difficult to continue practicing. Perhaps you've always wanted to learn to juggle! Find some instruction online and watch some videos to show you how you can get started, then practice for a few minutes each day. You will be surprised at how much better you feel emotionally and mentally as a result of exercising your brain in this way.

Daily Practice #5: Power Pose and Posture

Okay, so this daily practice should sound familiar, and it should be an easy daily practice to incorporate into your schedule. While the initial exercise doesn't last long, you will benefit a great deal from trying to maintain the posture and forming a habit of holding yourself in a way that boosts your confidence automatically. Let's review and try it.

Stand with your back straight and your shoulders back. Hold your chin parallel with the floor. Take it a bit further by putting a smile on your face. You will feel the difference, trust me!

Now, raise your arms above your head to form a V. Form fists with your hands, just like you see an athlete doing when he scores a touchdown or makes a basket in basketball. Hold your arms straight and maintain your posture. While you hold this pose, reflect on how this makes you feel. How did your mood change in just the last few seconds? Do you feel any different? How is your confidence level?

Instead of dismissing this exercise right away, challenge yourself to think about your posture as you go about your day. See how it make you feel and whether or not it alters your mood in any way. Most of all, see how it affects your confidence. You don't have to raise your arms up in a victory pose at work, but I highly recommend keeping some kind of reminder at your desk to help you remember to sit up straight and tall. This is not only much better for your back, but it's been shown to help confidence level as well as productivity. It is easier to feel in control and top of a heavy workload when your body exudes confidence. As your body reflects a position of confidence, it will infiltrate your mind until your attitude reflects the same confidence.

You can practice smiling as a way to improve your mood as well. Make it your goal to offer people smiles as you walk by them. You will most likely receive a warm smile in return, and it may even prompt a pleasant conversation, opening up possibilities for connection that may not have existed before.

Daily Practice #6: Practice Failure

This one might sound a little odd to you. What do I mean by practice failure? Wouldn't this just make you better at…failing? The answer is, surprisingly, no. And I will explain why.

Failure is an inevitable part of life. We fail a few times when we first learn to walk, to ride a bike, or climb a tree. We fail a few times when we first start trying to talk to individuals we're interested in dating, and we fail on first dates...at least, a lot of us do.

We need to prepare ourselves for the inevitable failure that comes with trying something new and challenging. It's when we are unprepared for how to handle failure that we crash and burn. Preparing yourself for what it might feel like to fail at a certain task will make the actual experience less devastating. When we set up expectations that are too high or too demanding, we set ourselves up for catastrophic failure. For example, if you had pressured yourself to adopt one new habit every single week without thinking about what this entails and how it might not be possible, you might have been devastated at the end of that first week, leading to frustration and perhaps even giving up on trying again. We've got to get to know ourselves as well as our limits. Get to know not only your strengths, but your weaknesses. Form a plan for a couple of different ways to approach a problem should you find that the first way is not the best. This will make you smarter and more ready for whatever comes around the corner.

You don't have to necessarily fail every single day as a way to practice failure, but a good daily practice is to wrap your mind around the possibility of failure, even in the simplest of tasks, then walk yourself through how it wouldn't be a big deal because you can always try again. Get your brain used to the possibility, and it won't so much of a surprise when those failures come around. Put a positive spin on failure by looking at it this way: You've found one more way *not* to do something! You can now avoid making the same mistakes in the future.

Daily Practice #7: Transform Your Morning

This daily practice will look different for each person depending on how their days are set up, but there are some general guidelines you can follow to optimize your morning for productivity, confidence and attitude throughout the rest of your day.

Does your current morning routine look something like this?—You hit the snooze once, twice…maybe three times. You roll out of bed, throw on some clothes, and head to the bathroom to brush your teeth as you think about how much you don't want to go to work that day. After brushing your teeth, you fix your hair for a second or two then fill a travel mug full of coffee before rushing out the door, just barely on track to get to work on time. Sound familiar?

If it does, I'm sure you're aware that this might not be the most ideal way to start your day. Our attitudes and moods at the very beginning of a day has a powerful influence on how the rest of our days are going to go. If we wake up in a bad mood, just sure that it's going to be a bad day...odds are, that's exactly how it's going to turn out. This happens because of the principle that we "see what we want to see." If we start out the day with a pessimistic attitude, we are going to go through our days seeing only the bad and the unlucky moments. But the truth is that this doesn't have to be our reality. We do have a choice...and it begins with redesigning your morning.

If you don't have much time in the morning and it stresses you out, then your first challenge is going to be to simply make more time for yourself in the morning by going to bed earlier and setting your alarm to go off a little earlier. A half hour is all you need, but you'll have to really put in the effort at first to change up your routine. Your body is going to scream at you to just lay there for another 30 minutes like usual, but I promise you, once you start to adopt this new morning ritual, you won't want to go back.

The idea is to set aside enough time for you to comfortably wake, have some kind of breakfast, then sit down for some personal "quiet time." The only requirement for quiet time is that you aren't online browsing the internet or any media sites. Take this time to read for a few minutes, start a new book you've been meaning to. Or, you could use this time to get in a mindfulness meditation session or practice that new skill you've decided to learn. Whatever it is, the idea is that it will put you in a calm, peaceful place before the chaos of the day takes over. Just a few minutes a day like this in the morning will make a big difference in your overall attitude, alertness, and confidence. And it won't take long to see this effect in action. Don't take my word for it. Try it!

Daily Practice #8: Pamper Yourself Once in a While

You may not have time every single day to do something extravagant, like a day at the spa or a new haircut or shopping spree, but you can take a few minutes each day to remind yourself of your reward as you earn it by conquering your daily habit-forming tasks.

Decide what you love best when it comes to pampering. We talked earlier about the power of positive reinforcement. You will be using the same concept here. Plan a day that will be all about you, when you can take it easy and relax, maybe treat yourself to a good meal at one of your favorite restaurants or take your partner out for a fun date night. Whatever you plan to do, make it a reality by setting down a hard plan and sticking to it once it comes around. Don't talk yourself out of it when the time comes with excuses like, Gee, I should save the money, or, I don't know if I worked hard enough for this. Don't let these discouraging voices keep you from rewarding yourself. Remember, you don't have be ready for the marathon after your first day of training. It takes consistent practice and dedication to accomplish your life goals. Treating yourself will give you something concrete and short-term to look forward to, and you'll be looking forward to the next treat as soon as you're done!

Daily Practice #9: Sleep!

This may seem like an obvious one, but you may be shocked at how many people do not get nearly enough sleep each night. When we don't get enough sleep, the body has to function on something called a sleep deficit, which is never made up for until we get that sleep that the body needs.

Many vitally important things happen when we sleep, and it is especially important for our brains. Set aside time to give yourself a solid 8 hours of sleep each and every night.

Daily Practice #10: Write

You knew this was coming. Keep up with the journaling! There is nothing more motivating than being able to look back and review your progress. See how far you've come and take notes on how your progress makes you feel. Just reading about past victories will automatically boost your confidence and keep you going for the rest of the day.

Conclusion

Conquering your bad habits through mind hacking to make way for new ones is one of the most important and life-changing endeavors a human being can pursue. From the time we are very young, we are programmed to feel and think a certain way until a point at which we begin to think for ourselves. Friends, family and authority figures can have a large impact on how we think and on the habits we cultivate, and these habits may continue for the rest of our lives.

When we begin to think and experience life on our own, we inevitably pick up some not-so-good habits along the way. These habits tend to develop slowly over time so that we do not even know they are a problem until they seem out of control. But these don't have to out of your control. Through the tips and strategies you've learned about in this book, you have the tools necessary to begin addressing these bad habits one by one before building up new, healthier habits in their place. These new habits are personal to you, and you should never feel pressured to adopt another's lifestyle out of obligation or pressure. Only you know what truly makes you happy and what you truly want out of life, and no one can take that from you.

You've learned the foundation of what mind hacking is all about as well as how our emotions can sometimes have a big impact on our behaviors. Because the feeling brain is quicker than the thinking brain, it can be easy to let yourself lose control and let emotions take the wheel in your mind. But the good news is that it doesn't have to be that way. You can exercise control over your emotions and behaviors through a proactive approach to thought.

After learning how emotion translates into behavior through some of the most outrageous and fascinating social experiments ever conducted, you were introduced to the principles of neuroplasticity. Neuroplasticity is what makes it possible for you to change your brain and redesign what you want your life to look like through cultivating new, better habits.

Even the most powerful emotions can be addressed through a consistent routine of mind-altering practices. You learned all about some of the most negative and hindering emotions a person can feel, including anxiety, worry and overthinking. You also learned how you can take control of these emotions by redirecting your negative thought trajectories toward a more positive pattern of thought.

Step 1 in the mind hacking process took you through the difficult process of identifying those negative outward influences which may be holding you back. After identifying and removing these negative influences from your life, you were ready to move forward to step 2.

Step 2 walked you through how to vividly portray your personal life goals through the practice of visualization. After creating a clear picture in your mind of where you were going, it was time to start taking your first steps.

Step 3 in the mind hacking process was all about setting small goals and taking small, manageable steps to gradually realize your goals. To keep you going throughout this process, you learned the fundamentals of meditation and how you can use this practice to strengthen your resolve to keep moving forward.

Finally, we introduced 10 effective daily practices to help you cultivate a sense of strength and self-confidence. As you continue on in your journey, remember to periodically reflect on your accomplishments and consider sharing your journey with others so that they might also have the opportunity to mind hack their way to success and happiness.

www.ingramcontent.com/pod-product-compliance
Lightning Source LLC
Chambersburg PA
CBHW031149020426
42333CB00013B/580